Ginger Cultivation and Its Antimicrobial and Pharmacological Potentials

Edited by Haiping Wang

Published in London, United Kingdom

IntechOpen

Supporting open minds since 2005

Ginger Cultivation and Its Antimicrobial and Pharmacological Potentials
http://dx.doi.org/10.5772/intechopen.83688
Edited by Haiping Wang

Contributors
Neeta Shivakumar, Vijayan Alavoor Keloth, Jebasingh Tennyson, Meenu Gupta, Gloria Otunola, Anthony Afolayan, Fernando Almeida De Souza, Amanda Mara Teles, Kátia Da Silva Calabrese, Ana Lucia Abreu-Silva, Adenilde Nascimento Mouchrek, Suwijiyo Pramono, Haiping Wang, Fatai Oladunni Balogun, Temitayo Esther Adeyeoluwa, Anofi Ashafa, Elias Sowley, Frederick Kankam

First published in London, United Kingdom, 2020 by IntechOpen
IntechOpen is the global imprint of INTECHOPEN LIMITED, registered in England and Wales, registration number: 11086078, 7th floor, 10 Lower Thames Street, London, EC3R 6AF, United Kingdom
Printed in Croatia

British Library Cataloguing-in-Publication Data
A catalogue record for this book is available from the British Library

Additional hard and PDF copies can be obtained from orders@intechopen.com

Ginger Cultivation and Its Antimicrobial and Pharmacological Potentials
Edited by Haiping Wang
p. cm.
Print ISBN 978-1-83880-029-1
Online ISBN 978-1-83880-030-7
eBook (PDF) ISBN 978-1-83880-407-7

For EU product safety concerns:
IN TECH d.o.o., Prolaz Marije Krucifikse Kozulić 3, 51000 Rijeka, Croatia, info@intechopen.com or visit our website at intechopen.com.

Meet the editor

Haiping Wang is a full-time research scientist and Professor at Department of Vegetables Germplasm, Institute of Vegetables and Flowers, Chinese Academy of Agricultural Sciences (IVF-CAAS). His research interest is research on vegetable genetic resources to collect germplasm for preservation of the diversity of the Midterm Gene-Bank of Vegetables Genetic Resources in China. Research on the genetics of garlic and ginger and breeding is conducted to improve the crop for growers and consumers. His key areas of interest include garlic, ginger, radish, and cucumber genetics and development of genomic tools, genetic improvement of garlic disease resistance, garlic diversity and origins, and human nutritional quality and flavor of both garlic and ginger. Outreach activities include interaction with garlic and ginger production and with consumers. Dr. Wang is the author and co-author of 44 publications in scientific journals and 11 publications in book chapters in Chinese or English. He has reviewed numerous publications for more than 10 international scientific journals.

Contents

Preface

It is well known that all plants in the family of ginger have economic value. They can be used as food additives, condiments, and medicines. Ginger is useful but requires a special habitat to survive. The scientific research results will be beneficial for the protection and utilization of ginger.

Ginger, as a bulbous plant, has been cultivated for a very long time. It is believed that ginger is native to southern China, Southeast Asia, and India. It was introduced to the Mediterranean in the first century, Japan in the third century, England in the eleventh century, and America in 1585. Today, it is widely grown all over the world as a major commercial spice crop. Ginger has also attracted more attention due to its antibacterial and pharmaceutical functions.

Although there have been many reports and research on ginger, it is necessary to have a comprehensive book covering this research. This book presents the latest findings and research covering such diverse fields as taxonomy and systems, molecular studies and phylogeny, phytochemistry and pharmacognosy, biodiversity and conservation, horticulture and hybridization, and all aspects of ginger biology. Authors present nine chapters of their work. In order to make it easier for readers to find the corresponding content, the book is divided into four sections. The "Introduction" section includes two chapters covering the basic information and a review of ginger. The "Cultivation and Improvement" section will help readers to understand how to grow and improve ginger. Some reviews and a number of events are presented in the sections "Antimicrobial Potentials" and "Pharmacological Potentials".

Due to the limited time and my academic level, there must be some deficiencies and mistakes in this book. Readers' corrections and criticisms are very welcome.

I would like to extend thanks to all authors who contributed to this book. My sincere thanks go to the author service manager, Ms. Marina Dusevic. She has walked me through all the stages of the book processing.

Haiping Wang
Institute of Vegetables and Flowers,
Chinese Academy of Agricultural Sciences,
Beijing, China

Section 1

Introduction

Chapter 1

Introductory Chapter: Studies on Ginger

Haiping Wang

1. Introduction

Ginger (*Zingiber officinale* Roscoe), as a bulbous plant, has been cultivated for a very long time. It is believed that ginger is native to southern China, Southeast Asia, and India. It was introduced to the Mediterranean in the first century, Japan in the third century, England in the eleventh century, and America in 1585 [1]. It is now widely cultivated in tropical and subtropical regions of the world but mostly in Asia and Africa. The total harvest area of ginger in the world is more than 21,000 hac with a total production of more than 200,000 tons and an average yield of 10,000 kg per hac. Ginger is mainly used as spice and flavour agent for food. The characteristic fragrance and flavour of ginger are the result of volatile oils, primarily consisting of zingerone, shogaols and gingerols as the major pungent compound. A lot of studies have been carried out to discover the miracle of this plant. The book will cover the history of ginger in cultivation, therapeutic benefits, modern cultivation and production, varieties and breeding.

2. Botanical character

The leafy stems of ginger grow about a metre high. The leaves are 6–12 inches (15–30 cm) long, elongate, alternate in two vertical rows and arise from sheaths enwrapping the stem. The flowers are in dense cone-like spikes about 1-inch thick and 2–3 inches long that are composed of overlapping green bracts, which may be edged with yellow. Each bract encloses a single, small, yellow-green and purple flower. The consumed portion of the ginger plant is the rhizome, often called "ginger root", although it is not an actual root. The rhizome is the horizontal stem of the plant that sends out the roots [2].

3. The history of ginger cultivation

Ginger, an herbaceous perennial plant of the family Zingiberaceae, probably native to southeastern Asia [3], or its aromatic, pungent rhizome, is used as a spice, flavouring, food and medicine. Its generic name *Zingiber* is derived from the Greek zingiberis, which comes from the Sanskrit name of the spice, singabera. Its use in India and China has been known from ancient times, and by the first century, traders had taken ginger into the Mediterranean region. By the eleventh century, it was well-known in England. The Spaniards brought it to the West Indies and Mexico soon after the conquest, and by 1547 ginger was being exported from Santiago to Spain [4].

4. Ginger is mainly used as spice and flavouring agent for food

Ginger is well-known as spice and flavouring agent for food [5]. Ginger is used in cooking, in various forms such as immature ginger, mature fresh ginger, dry ginger, ginger oil, ginger oleoresin, dry-soluble ginger, ginger paste and ginger emulsion. It is rich in secondary metabolites, namely, the oleoresins, contributing widely the pungency and flavours. Ginger (*Zingiber officinale* Rosc.), which belongs to the family Zingiberaceae, is an important tropical horticultural plant and an important spice crop used in various medicinal and culinary preparations [6]. Besides, ginger is very popular in the food industry as an additive to ginger ale, candies, pastries and cakes [7]. Its uses, of course, aren't confined to food preparation.

5. Ginger consumption is known for its health benefits

Ginger is an excellent source of several bioactive phenolics, including non-volatile pungent compounds such as gingerols, paradols, shogaols and zingerones. Ginger is also used in traditional oriental medicine (Ayurvedic, Chinese and Unani systems of medicine) since antiquity (>2500 years) to treat different diseases that include rheumatoid arthritis, sprains and muscular aches, sore throats, nausea, constipation and indigestion, fever, infectious diseases and helminthiasis [8, 9]. It is particularly valued in medicine as a carminative and stimulant to the gastrointestinal tract. Ginger consumption is known for its health benefits and widely known to be used in Ayurvedic formulations and Chinese medicine. It is stimulative in nature and helps in relieving indigestion, stomach ache, diarrhoea and nausea. It is widely used to cure common cold, cough and congestion. Clinical studies have demonstrated it to be antiemetic, antiulcer, anti-platelet, anti-inflammatory and antioxidant in nature [10, 11]. Ginger has many uses as home remedies and can be used to help arthritis, diarrhoea, flu, headache, heart and menstrual problems, diabetes, stomach upset and motion sickness. Wide studies have been taken up involving ginger to cure complex diseases such as cancers to the chronic conditions of migraines.

6. Germplasm and varieties

There are many genetic resources which were collected and preserved in the word. Zingiberis family includes about 50 genera and 1300 species of ginger which are known to exist worldwide [12–15]. They occur in different parts of the world, namely, Japan, Australia, Haiti, Bangladesh, Jamaica, Sri Lanka and Nigeria. However, most of the varieties used in commercial production were reported from India and China [16, 17]. Several cultivars of ginger are grown in different ginger-growing areas in India, and they are generally named after the localities where they are grown [3]. Some of the prominent indigenous cultivars of ginger grown in India are Himachal, Maran, Kuruppampady, Wayanad, Varadha, etc. Exotic cultivars such as Rio de Janeiro have also become very popular among cultivars [3]. Maran, Nadia, Karakkal and Rigodi are suited for high dry ginger. Varieties like Ernadn Chrnad, China and Rio de Janeiro provide high oleoresin content. Sleve local, Narasapattam and Himachal are suited for high volatile oil. Rio de Janeiro, China, Wayanad, Maran and Varadha are suited for green ginger. The production of many very popular varieties used in China is exporting to other countries [18, 19]. The most famous varieties includes Laiwu ginger, Tongling ginger, Pinghu ginger, Laifeng ginger, etc. Due to vegetatively propagated characteristic, in vitro techniques, namely, micropropagation techniques [20], somatic embryogenesis [21], somatic hybridization[21], germplasm conservation, transgenics and mutation breeding, are mostly used [21, 22].

7. Production and disease arrangement during the cultivation

Ginger crop is affected by insect pests, pathogenic and non-pathogenic diseases [23] and severely by various pathogenic diseases of viral, bacterial, fungal and nematode origin, which reduces its potential yields drastically [24]. Among the various diseases, soft rot, yellows, *Phyllosticta* leaf spot, storage rot, bacterial wilt, mosaic and chlorotic fleck are important. Therefore, the selection of healthy seed rhizomes has been found as an effective control measure for the disease. Efficacy of a variety of chemicals has been evaluated for the management of this disease by different workers, and they have found very promising effect of different chemicals against the disease.

8. Biological management is a prosperous way to control ginger diseases

Application of a mixture of biological bacteria could be very promising to increase rhizome production [25–28]. By using resistant or less susceptible cultivars of ginger, the disease can be managed to a great extent [17].

Author details

Haiping Wang
Institute of Vegetables and Flowers, Chinese Academy of Agricultural Sciences, Beijing, China

*Address all correspondence to: wanghaiping@caas.cn

References

[1] Langner E, Greifenberg S, Gruenwald J. Ginger: History and use. Advances in Therapy. 1998;**15**(1):25-44

[2] Ravindran P, Babu KN, Shiva K. Botany and crop improvement of ginger. In: Ginger. Boca Raton, USA: CRC Press; 2016. pp. 35-106

[3] Xizhen A, Jinfeng S, Xia X. Ginger production in Southeast Asia. In: Ginger. Boca Raton, USA: CRC Press; 2016. pp. 261-298

[4] Okwuowulu P. Ginger in Africa and the Pacific Ocean Islands. In: Ginger. Boca Raton, USA: CRC Press; 2016. pp. 299-324

[5] Bartley JP, Jacobs AL. Effects of drying on flavour compounds in Australian-grown ginger (*Zingiber officinale*). Journal of the Science of Food and Agriculture. 2000;**80**(2):209-215

[6] Srinivasan K. Ginger rhizomes (*Zingiber officinale*): A spice with multiple health beneficial potentials. PharmaNutrition. 2017;**5**(1):18-28

[7] Malu S, Obochi G, Tawo E, Nyong B. Antibacterial activity and medicinal properties of ginger (*Zingiber officinale*). Global Journal of Pure and Applied Sciences. 2009;**15**(3-4)

[8] Masuda Y, Kikuzaki H, Hisamoto M, Nakatani NJB. Antioxidant properties of gingerol related compounds from ginger. BioFactors. 2004;**21**(1-4):293-296

[9] Shukla Y, Singh M. Cancer preventive properties of ginger: A brief review. Food and Chemical Toxicology. Boca Raton, USA. 2007;**45**(5):683-690

[10] Ali BH, Blunden G, Tanira MO, Nemmar A. Some phytochemical, pharmacological and toxicological properties of ginger (*Zingiber officinale* Roscoe): A review of recent research. Food and Chemical Toxicology. 2008;**46**(2):409-420

[11] Ghasemzadeh A, Jaafar HZ, Rahmat A. Antioxidant activities, total phenolics and flavonoids content in two varieties of Malaysia young ginger (*Zingiber officinale* Roscoe). Molecules. 2010;**15**(6):4324-4333

[12] Ismail NA, Rafii M, Mahmud T, Hanafi M, Miah G. Genetic diversity of torch ginger (*Etlingera elatior*) germplasm revealed by ISSR and SSR markers. BioMed Research International. 2019;**2019**:5904804

[13] Kumar A, Kapoor C, Rahman H, Karuppaiyan R, Rai S, Denzogpa R. Multivariate analysis of ginger (*Zingiber officinale* Rosc.) germplasm of North Eastern India. Indian Journal of Genetics and Plant Breeding. 2016;**76**(2):221-223

[14] Blanco EZ, Bajay MM, Siqueira MVBM, Zucchi MI, Pinheiro JB. Genetic diversity and structure of Brazilian ginger germplasm (*Zingiber officinale*) revealed by AFLP markers. Genetica. 2016;**144**(6):627-638

[15] Das A, Gaur M, Barik D, Subudhi E. Genetic diversity analysis of 60 ginger germplasm core accessions using ISSR and SSR markers. Plant Biosystems. 2017;**151**(5):822-832

[16] Prasath D, Kandiannan K, Srinivasan V, Leela N, Anandaraj M. Comparison of Conventional and Transplant Production Systems on Yield and Quality of Ginger (*Zingiber officinale*). Indian Journal of Agricultural Sciences. 2018;**88**(4):615-620

[17] Nybe EV, Raj NM. Ginger production in India and other South

Asian countries. In: Ginger. Boca Raton, USA: CRC Press; 2016. pp. 231-260

[18] Binghong X, Chunyan L, Xinlan X, Bo C, Caixia A, Shiquan L, et al. Studies on collection breeding and application of *Zingiberaceae* plants wild resources in China. Journal of Plant Sciences. 2018;**6**(5):179-184

[19] Huang C, Zhou Q, Gao S, Bao Q, Chen F, Liu C. Time-domain nuclear magnetic resonance investigation of water dynamics in different ginger cultivars. Journal of Agricultural and Food Chemistry. 2016;**64**(2):470-477

[20] Inden H, Asahira T, Hirano A. Micropropagation of ginger. In: Symposium on High Technology in Protected Cultivation. Vol. 230. 1988. pp. 177-184

[21] Musfir Mehaboob V, Faizal K, Thilip C, Raja P, Thiagu G, Aslam A, et al. Indirect somatic embryogenesis and Agrobacterium-mediated transient transformation of ginger (*Zingiber officinale* Rosc.) using leaf sheath explants. The Journal of Horticultural Science and Biotechnology. 2019;**94**:753-760

[22] Loyola-Vargas VM, Ochoa-Alejo N. Somatic embryogenesis. An overview. In: Somatic Embryogenesis: Fundamental Aspects and Applications. Berlin, Germany: Springer; 2016. pp. 1-8

[23] Yuan Y, Gao M. Characteristics and complete genome analysis of a novel jumbo phage infecting pathogenic *Bacillus pumilus* causing ginger rhizome rot disease. Archives of Virology. 2016;**161**(12):3597-3600

[24] Dohroo N. Diseases of ginger. In: Ginger. Boca Raton, USA: CRC Press; 2016. pp. 325-360

[25] Rai M, Ingle AP, Paralikar P, Anasane N, Gade R, Ingle P. Effective management of soft rot of ginger caused by *Pythium* spp. and *Fusarium* spp.: Emerging role of nanotechnology. Applied Microbiology and Biotechnology. 2018;**102**(16):6827-6839

[26] Bhattarai K, Pokharel B, Maharjan S, Adhikari S. Chemical Constituents and Biological Activities of Ginger Rhizomes from Three Different Regions of Nepal. Journal of nutritional dietetics & probiotics. 2018;**1**(1):1-12

[27] Debata D, Sethy A, Panda D, Sarangi P. Management of rhizome rot of ginger. Environment and Ecology. 2019;**37**(1):97-100

[28] Anil T, Nisha T, Dohroo N. Rhizome rot of ginger-management through non-chemical approach. International Journal of Plant Protection. 2017;**10**(1):140-145

Chapter 2

Utilisation and Functional Components Evaluation of Ginger

Suwijiyo Pramono

Abstract

Ginger is a Zingiberaceae plant having different purposes in the community and industry. The important parameters of quality of ginger are the functional components so the aims of this chapter are to review the utilisation of ginger in the community and industry and to evaluate the functional components of ginger and its products. Ginger (*Zingiber officinale* Roscoe) has at least three types, i.e. big ginger, small ginger and red ginger. Fresh, dried and preserved ginger and also its extract, oleoresin and volatile oil were considered as basic products of the utilisation of ginger. Different formulas have been developed for drinks, culinary purposes, flavouring desert and herbal medicines. In folk medicines, ginger is used as remedy for warming body, gastritis and fracture condition. Based on scientific researches, ginger has been developed as anti-emetic, anti-inflammatory, analgesic and anti-influenza. Evaluation of chemical constituents of ginger and its products can be done qualitatively for authentication and quantitatively for standardization. This chapter consists of the utilisation of ginger based on empirical and scientific data, and the functional components evaluation consisting of authentication and standardization.

Keywords: ginger, utilisation, functional components, authentication, standardization

1. Introduction

Ginger (*Zingiber officinale* Roscoe) is a Zingiberaceae plant having different purposes in the community and industry. People use its rhizome as spice, drink, or as a component of herbal medicines. People use fresh or dried rhizomes and preserved ginger. The scientific name of plant material is Rhizoma Zingiberis. Actually there are several genus *Zingiber* in Zingiberaceae family such as *Zingiber zerumbet*, *Z. amaricans*, *Z. aromaticum* and *Z. purpureum* and the name of plant material must completely refer to the name of species such as Rhizoma Zingiberis Zerumbeti and Rhizoma Zingiberis Purpurei, but especially for ginger, the name refers only to genus even though some Pharmacopoeias use the complete name: Rhizoma Zingiberis Officinalis [1]. There are three varieties of ginger, i.e. *Zingiber officinale* var. *officinale*, namely big white ginger or big ginger; *Zingiber officinale* var. *amarum*, namely small white ginger or small ginger; and *Zingiber officinale* var. rubrum, namely red ginger. In Indonesia, big white ginger is called jahe gajah (jahe: ginger, gadjah: elephant), small white ginger is called jahe emprit (jahe: ginger, emprit: small bird) (**Figure 1**).

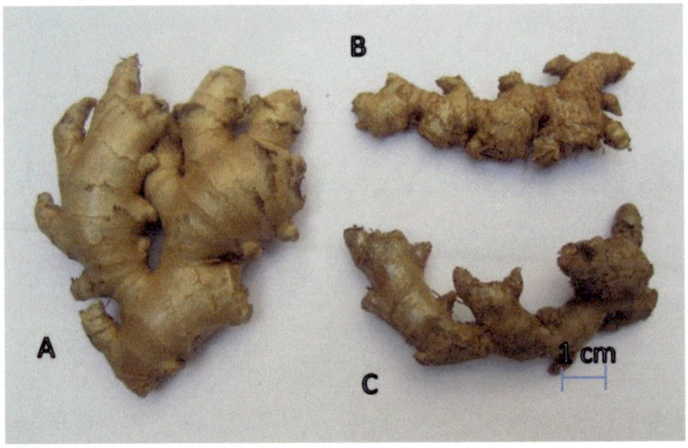

Figure 1.
Fresh ginger: (A) big ginger, (B) small ginger and (C) red ginger.

2. Products of ginger

2.1 Fresh ginger

In South-East Asia, fresh ginger is used in cooking as flavouring or as a vegetable and also to make ginger ale or other drinks. Fresh ginger is prepared from immature or mature rhizomes. Big ginger contains less fibre and is less odorous and less pungent than small and red ginger. Traditionally, people take three fingers of fresh ginger to grate and bandage a sprained leg. The decoction of fresh ginger is taken orally to cure nausea and vomitus [2]. Some traditional drinks often use fresh ginger than the dried one in order to keep the aroma. In the market, there are a lot of ginger products especially in instant dosage form with hot taste but lack of aroma. A remedy for headache caused by influenza is to use 10 g of grilled fresh ginger and 20 g of coconut sugar, boiled with 200 mL water and drink in warm condition. Javanese people have special drink, namely 'serbat', used for warming body, which consists of 5 g fresh ginger, 2 g lemongrass, 2 g clove, 1 g nutmeg, 1 g cinnamon, 2 g kaffir lime leaves, 5 g cubeba fruits and 100 g palm sugar, boiled with 100 mL water for 10 min and then drink 2 times a day, 1 glass each [3].

2.2 Dried ginger

Ginger occurs in horizontal, laterally flattened, irregularly branching pieces, 3–16 cm long, 3–4 cm wide, up to 2 cm thick, sometimes split longitudinally; pale yellowish buff or light brown externally, longitudinally striated, somewhat fibrous, branches known as 'fingers' arise obliquely from the rhizomes, are flattish, obovate and short, about 1–3 cm long; fracture, short and starchy with projecting fibres. Internally, the colour is yellowish brown, showing a separating the narrow cortex from the white stele. It contains and numerous scattered fibrovascular bundles, scattered on the whole surface. Ginger has the following characteristics: odour, aromatic; taste, pungent and aromatic; and colour, internally pale yellow to brown [4]. Dried ginger is used in the form of powder and is applied worldwide for domestic culinary purposes, and also extensively in the flavouring desserts. In folk medicine, dried-powdered ginger is used as component of remedies for certain indications. Powder of 1 g ginger, 1 g cardamom and 1 g cinnamon are mixed, divided into three sachets and taken orally after mixed with 200 mL boiled water, 3 times a day, one sachet each. This remedy is used for curing throat inflammatory [3]. A mixture of 1 g powdered ginger and 2 g rhubarb is divided into three sachets and then taken orally 3 times a day after treated with boiled water to reduce gastritis [3].

2.3 Preserved ginger

There are two methods in preparing preserved ginger. The first one is started by freezing the fresh ginger overnight and then peeling and cutting it into small pieces. Place water and ginger in a 3-quart pan. Remove the ginger and put the liquid back in the pot, add sugar and then boil until the liquid becomes syrup [5]. The second method begins by grating fresh ginger to obtain porridge. Press the porridge with addition small quantity of water and then filter with a funnel and tissue. The sap is evaporated until viscous and then white sugar or maltodextrin is added until crystallization. In Indonesia, this product is popularly named instant ginger product. There are a lot of these dosage form sold in the market with promotive indication [6].

2.4 Extract, oleoresin and essential oils of ginger

The fresh and dried rhizomes of ginger yield an essential oil ('ginger oil') and oleoresin ('ginger extract'). Ginger oil has the aroma and flavour of the spice but lack of pungency. It is used for flavouring beverages and in cosmetics, perfumes and pharmaceuticals. Ginger oleoresin has the aroma and flavour and pungency of the spice. It is more often used in pharmaceuticals. In the United States, the regulatory status 'generally recognized as safe' has been accorded to ginger (GRAS 2520), ginger oil (GRAS 2522) and extract/oleoresin (GRAS 2521/2523) [4].

Ginger has been used medicinally in Asia since ancient times, e.g. in China and India. In Indonesia, there is a term of indication that is not recognized in medical dictionary, i.e. 'masuk angin'. The term "masuk angin" is defined as a weakness condition of body with several symptoms such as flatulent, cool in sweat but high temperature of body, sleepy and pain in the muscles and the bones. The weakness condition of body may be caused by fattigue, decrease of immune system or influenza. The products contain ginger extract as the main component, and it is considered as carminative, stimulant of gastrointestinal tract, rubefacient and counterirritant. An example of famous product consists of Rhizoma Zingiberis, Herba Echinacea, Radix Valerian, Radix Panax ginseng, Fructus Foeniculi, Fructus Isorae, Semen Myristicae, Fructus Amomi, Folium Caryophylli and Herba Menthae Arvensis [6]. People consider that 'masuk angin' is similar to influenza. Furthermore, red ginger is considered having more potential as aphrodisiac in comparison to small and big white ginger.

2.5 Pharmacological effects based on utilisation

Pharmacologically, ginger has anti-inflammatory activity with the increase of arachidonic acid oxidation by inhibition of cyclooxygenase and 5-lipoxygenase, resulting in the synthesis of prostaglandin E and leukotriene B_4. An in vitro experiment showed that the aqueous extract of ginger inhibited cyclooxygenase and lipoxygenase and decreased prostaglandin and leukotriene. The in vivo experiment showed that oral treatment of rats with ginger extract reduced paw oedema. Furthermore, an artefact constituent of ginger, namely shogaol, reduced paw oedema of rats induced by carrageenan. Two labdane compounds isolated from ginger showed their ability as inhibitor of 5-lipoxigenase in in vitro experiment. Injection of sterile preparation of ginger extract to 113 patients suffering from rheumatic condition and back bone pain in China reduced the level of pain, nodules and inflammatory, and ameliorated the bone function [7]. Ginger extract showed peripheral antiemetic activity in dog but it did not act on central nervous system. This anti-emetic effect is caused by synergism between zingerone and shogaol. A clinical study showed that the oral treatment of 90 g powdered ginger was more effective in comparison to dimenhydrinate in reducing motion sickness symptom [8]. The oral treatment of 500 mg/kg BW water extract of ginger for 4 weeks decreased significantly blood cholesterol levels in rats but did not decrease triglyceride levels. Intraperitoneal treatment of this extract decreased blood cholesterol in rats [8].

A crossover design of a double-blind, randomized placebo-controlled study on 13 patients with history of motion sickness and circular vection showed that pretreatment of 1000 and 2000 mg water extract of ginger reduced significantly headache, nausea and plasma vasopressin [8]. Ginger also increased tonus and peristaltic of stomach [9]. A double-blind randomized trial without placebo compared with scopolamine, dimenhydrinate and other conventional drugs on 1489 voluntary subjects 2 h after pretreatment with 500 mg ginger before journey with car did not have problem of motion sickness, statistically similar to all tested conventional drugs [10]. A one double-blind placebo-controlled trial on women after gynaecologic surgery randomly treated with 1 g powdered ginger or 10 mg metoclopramide showed that patients with ginger treatment had less problem of vomiting [11]. From six double-blind randomized clinical trials (RCT) with a total of 675 voluntary patients and a prospective observational cohort study ($n = 187$), 4 of 6 RCT showed high significant difference between ginger treatments and placebo; 2 RCT showed that ginger treatments were more effective than those of Vit B_6 in managing nausea and vomiting. There was no report on the undesirable effect in pregnancy [12].

Based on empirical and scientific data until 1999, the World Health Organization divided ginger utilisation into three categories [13]:

Uses described in folk medicine, not supported by experimental or clinical data: to treat cataracts, toothache, insomnia, baldness, and haemorrhoids and to increase longevity.

Uses described in pharmacopoeias and in traditional systems of medicine: to treat dyspepsia, flatulence, colic, vomiting, diarrhoea, spas and other stomach complaints. Powdered ginger is further employed in the treatment of colds and flu, to stimulate the appetite, as a narcotic antagonist and as an anti-inflammatory agent in the treatment of migraine headache, rheumatic and muscular. In addition, Chinese Materia Medica mentioned the use of ginger in the treatment of abdominal pain due to cold from deficiency and stagnant blood; to warm the middle and expel cold: for warming the spleen and stomach both in condition of excess due to externally contracted cold, as well as cold from deficiency due to insufficiency of the yang qi; to warm the lungs and transform phlegm: for lung cold with expectoration of thin, watery, or white sputum; to warm the channels and stop bleeding: for cold from deficiency that may present with haemorrhage of various types, especially uterine bleeding. It is used in treating haemorrhage only if the bleeding is chronic and pale in colour, and is accompanied by cold limbs, ashen white face, and a soggy, thin pulse [14].

Uses supported by clinical data: the prophylaxis of nausea and vomiting associated with motion sickness, postoperative nausea, pernicious vomiting in pregnancy, and seasickness. However, the use of ginger for early pregnant woman must be careful because there were several experiments that showed teratogenic effect on rats.

In addition, based on scientific data published after 1999, utilisation of ginger has been developed as an analgesic and anti-inflammatory agent. Ginger extract has a beneficial influence on morphine analgesia and can be an efficacious adjunct for pain management [15]. Ginger oil (0.25–1.0 g/kg) inhibited significantly carrageenan-induced paw oedema, active as adjuvant arthritis, anti-inflammatory mediators-induced vascular permeability in rats [16]. An ethanolic extract of ginger (50 and 100 mg/kg BW) produced significant inhibition of carrageenan-induced rat paw oedema and a reduction in the number of writhing induced by acetic acid in mice [17]. Other potential benefits of ginger are antimicrobial, lowering blood pressure, lowering cholesterol, antiplatelet aggregation, chemopreventive agent, antioxidant and hypoglycemic properties [18]. However, the antiplatelet aggregation effect gives the possibility in enhancing blood dilution effect of acetosal, vitamin K, heparin and other blood dilution substances. Ginger essential oils were reported having appetite stimulant property in rats, so it will be contradictive with its lowering cholesterol effect. The relatively high dose of ginger can stimulate lacrimation due to its hot taste.

3. Evaluation

The use of traditional medicines in a country is part of the practice and culture for centuries. However, in general, herbal materials have not been supported by adequate standardization. In this respect, it is deemed necessary to establish standards of herbal material as reference to reach optimum health control. In general, pharmacopoeia is a good choice to be a reference for the evaluation of an herbal material such as ginger. If an herbal material complies to the standard, it will give reproducibility of product to be marketed. Evaluation of herbal material consists of qualitative and quantitative assessment. The qualitative assessment is applied for identification and authentication, while the quantitative assessment is applied for standardization. Authentication of ginger can be done by evaluation of description, microscopic data and chromatography. The last one can be thin-layer chromatography, high-performance liquid chromatography or other qualitative methods. Standardization of ginger consists of loss on drying, total ash, acid-insoluble ash, water-soluble extract, ethanol-soluble extract and chemical content. The last one can be determined by spectrophotometry or chromatography analysis.

4. Description

The sliced thin rhizome has a short-branch edge and reverse egg shape, is usually 3–4 cm in length and has 1–6.5 mm thickness, and is yellowish white coloured. Outer part is yellowish brown. Cross-sectional cut shows narrow cortex, with the depth of less than a third from the radius and endoderm. Xylem vessel is spread and greyish; oil globules are yellowish smaller spots. It has strong characteristic odour and spicy taste [1] (**Figure 2**).

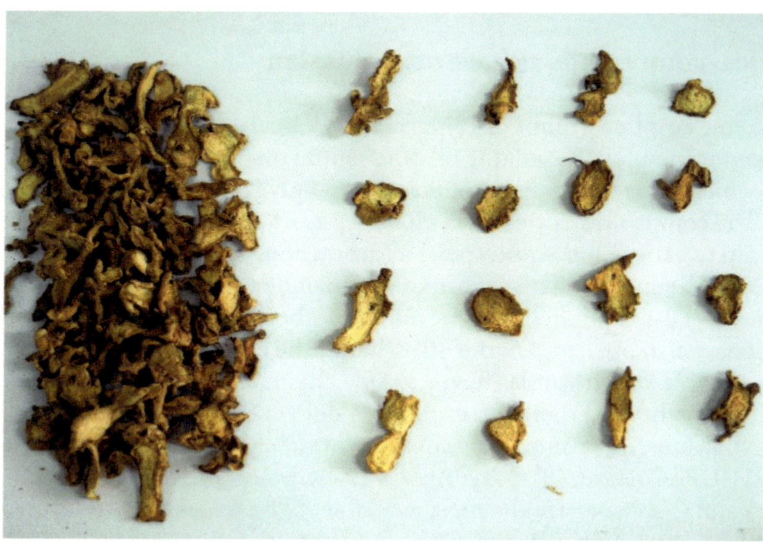

Figure 2.
Dried ginger rhizome.

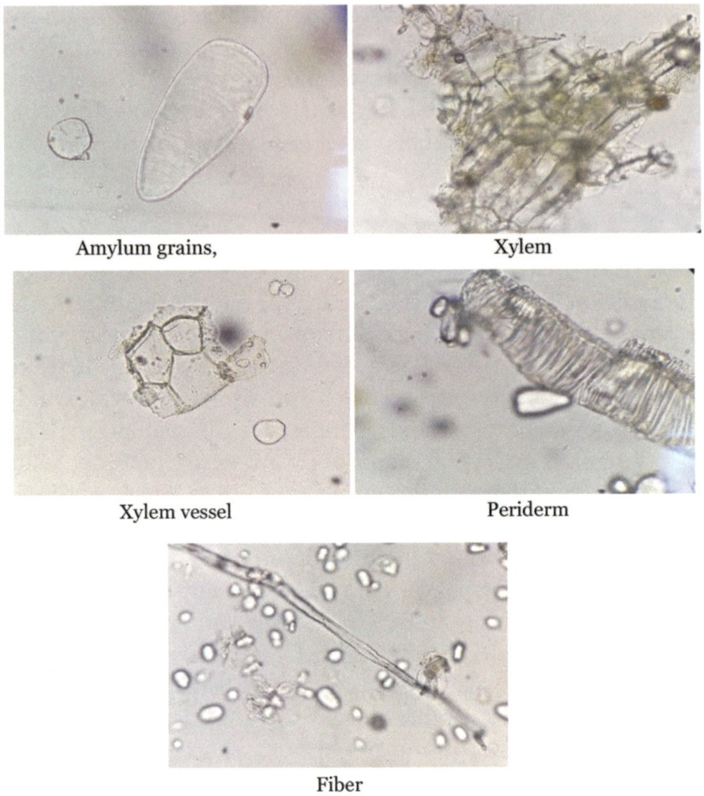

Figure 3.
Fragments of ginger powder.

5. Microscopic

Identification fragments are amylum, xylem, xylem vessel, periderm and fibers [1] (**Figure 3**).

6. Chemical compounds as base of evaluation

Ginger contains two groups of important chemical constituents: 1 g volatile oil and pungent principle. The volatile oil or essential oil gives the odour of the plant material, while the second one is not volatile and gives the pungent taste of the rhizome. The composition of volatile oil varies as a function of geographical origin, but the main constituent sesquiterpene hydrocarbons (responsible for the aroma) seems to remain constant. These compounds include (−) zingiberene, (+) ar-curcumene, (−) β-sesquiphellandrene and β-bisabolene. Monoterpene aldehyde and alcohol are also present. On the other hand, the chemical components of pungent principle are gingerols (having a side chain with 7–10, 12, 14 or 16 carbon atoms, respectively) and their corresponding dehydration products which are known as shogaols [19]. Among the components of the pungent principle of ginger, 6-gingerol (IUPAC name: (S)-5-hydroxy-1-(4-hydroxy-3-methoxyphenyl)-3-decanone) and 6-shogaol are the most available in the market. Thin-layer chromatography (TLC) and high-performance liquid chromatography (HPLC) are suitable for analysis of pungent principles, while gas chromatography-mass spectroscopy is suitable to identify the components of volatile oil. Even though if the sample to be analysed is in the form of ethanolic extract, the obtained data will contain components of volatile oil and those of pungent principle.

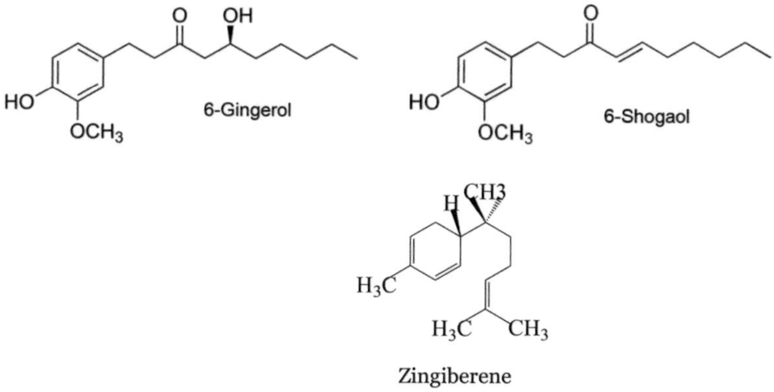

7. Thin-layer chromatography

If one of the standard references (zingiberene, 6-gingerol or 6-shogaol) is available, use it for TLC analysis. If it is not available, use eugenol as standard reference.

Carry out thin-layer chromatography (TLC) using the following parameters:

Mobile phase: toluene—ethyl acetate (93:7 v/v); stationary phase: silica gel 60 F_{254}; test solution: 10% ginger powder in ethanol; standard solution: 1.0% eugenol in ethanol; spotted volume: apply separately 3 μL test solution and 1 μL standard solution in the plate; detection: anisaldehyde-sulphuric acid, dry at temperature 100°C for 5–10 min (**Figure 4**).

This procedure of TLC analysis uses eugenol as standard reference because it is not easy to obtain shogaol as a marker substance. The correct evaluation is TLC with zingiberene, gingerol or shogaol as a standard reference and Rf will be used instead of Rx. Furthermore, the use of reagent for visualization such as anisaldehyde-sulphuric acid is lack of reproducibility. If the constituents of plant material have fluorescence under UV light such as those from ginger, it will be better to perform TLC without reagent for visualization. The following figures represent TLC profile of the three varieties of ginger with same system of TLC such as mentioned above.

Mobile phase: toluene—ethyl acetate (93:7 v/v); stationary phase: silica gel 60 F_{254}; test solution: 10% ginger powder in ethanol; standard solution: 1.0% 6-shogaol in ethanol; spotted volume: apply separately 3 μL test solution and 1 μL standard solution in the plate; detection: UV_{366} (**Figure 5**).

The quantitative evaluation can be done by TLC-densitometry if shogaols or gingerols are available in laboratory. If not, UV/vis spectrophotometry can be used to determine total phenol using Folin-Ciocalteu method. In this method, eugenol can be used as a standard reference. It must be noted that TLC method is the most recommended method for phytochemical screening and not only for the colour and precipitation observation in the tube. As we know that the colours of extract to be tested are generally green or light chocolate. It gives confusion to the colour produced by reaction occurs. In the colour reaction test of flavonoid for exemple, the produced colour in the tube is yellow. If it is not intensively appears it will be difficult to justify when the extract solution in the tube is green or light chocolate. There were some publications reporting the existence of alkaloid in ginger according to colour reaction in tube. After verification by TLC method, the result was negative, so the colour reaction in tube was not true and we called it as a false

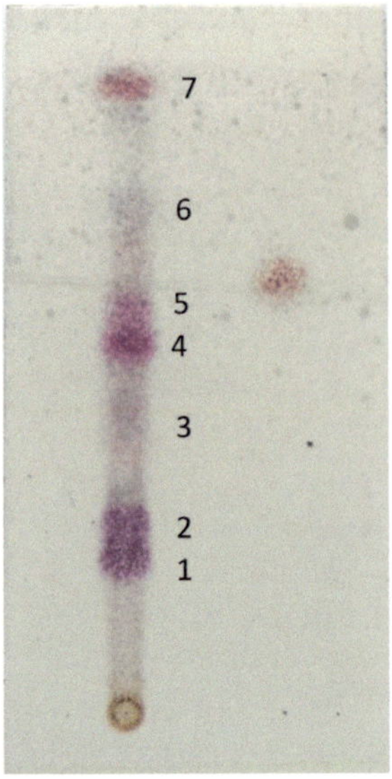

S: Sample Ginger

R: standard reference eugenol

Rf standard reference eugenol 0,67

Rx 1. 0,37

Rx 2. 0,45

Rx 3. 0,73

Rx 4. 0,82

Rx 5. 0,91

Rx 6. 1,15

Rx 7. 1,43

Figure 4.
TLC profile of ginger.

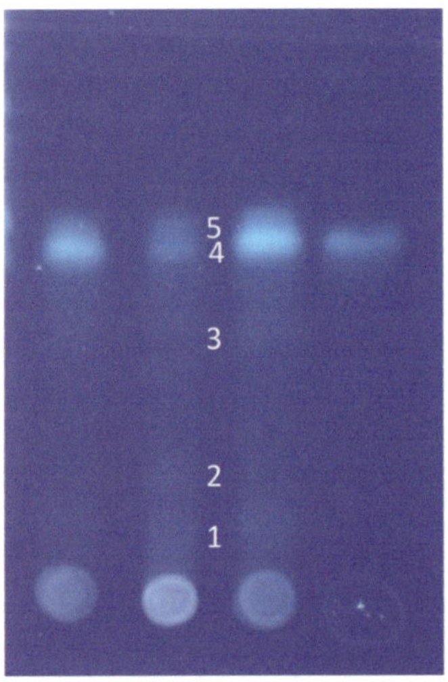

Rf reference standard shogaol: 0.70

R_f 1. 0.10
R_f 2. 0.25
R_f 3. 0.55
R_f 4. 0.70
R_f 5. 0.75

Figure 5.
TLC profile of 3 ginger varieties: (A) small ginger, (B) big ginger, (C) red ginger and (D) 6-shogaol.

positive reaction. According to chemotaxonomic approach, the existence of alkaloid in Zingiberaceae family is very rare. The similar case is sometimes found in the colour reaction of steroid. The appearance of red-pink colour in green solution is not easy to be detected; hence, it will be better and clear if we use TLC methods.

8. High-performance liquid chromatography

There are several publications concerning the HPLC analysis of ginger. Kawakishi et al. used Develosil ODS-5 column (8 × 250 mm), methanol-water-acetic acid (80:20:1, v/v/v) mobile phase at flow rate of 3 mL/min; detection at 254 nm [20]. Schwertner and Rios analysed 6-gingerol, 8-gingerol, 10-gingerol and 6-shogaol in ginger-containing dietary supplements, spices, teas and beverages on C-8 reverse phase column at 282 nm [21]. A simple HPLC method for analysis 6-gingerol in multiple shoot cultures of ginger was performed by Cafino using an ODS column and the best mobile phase was found to be methanol-water (90:10, v/v) with retention time observed at 3.8 min [22].

9. Gas chromatography-mass spectrophotometry

Ginger contains two main constituents, volatile oil and pungent substances. The first one gives the specific aroma of ginger, while the second one expresses the pungent taste of ginger. The volatile oil can be isolated by distillation method and then Gas chromatography-mass spectrophotometry can be used for determination of the components.

In order to evaluate the different components in ginger volatile oil, the peaks of chromatogram can be determined, and their chemical structure by comparing the recorded fragmentation with library. Furthermore, quantitative composition of volatile oils can be seen from the values of area versus high of peak (A/H).

Table 1 represents the chemical composition of volatile oils of three varieties of ginger. There is no significant difference between small and big ginger. Some constituents of small ginger such as nerol, E-citral, endobornyl acetate and neryl acetate are not present in both the other variety of ginger. On the contrary, a component of big ginger, namely, geranyl acetate is not present in small ginger, while interestingly volatile oil of red ginger contains substances totally different with those of small and big ginger. These data are in line with the odour of those three ginger. The aroma of small ginger is stronger than big ginger, while the aroma of red ginger is totally different.

It can be seen in **Figure 6** that all peaks of red ginger essential oil are recorded after 8.9, and there is no overlap with big and small ginger that also can be seen in **Table 1.**

Retention times	Dried small ginger (area/high)		Dried big ginger (area/high)		Dried red ginger (area/high)
8.979	Z-citral	6.04	Z-citral	6.49	
9.261	Nerol	2.64			
9.342	E-citral	1.41			
9.527	Endobornyl acetate	1.26			
10.296			Geranyl acetate	1.73	
10.298	Neryl acetate	2.26			
11.165	Ar-curcumene	1.56	Ar-curcumene	1.24	
11.280	Zingiberene	1.27	Zingiberene	1.48	
11.325			Farnesene	1.14	
11.385	cis-Farnesol	1.18	cis-Farnesol	1.24	
11.493	Beta-sesquiphellandrene	1.22	Beta-sesquiphellandrene	1.29	
11.649			Elemol	1.41	
12.072	Zingiberenol	1.47			
12.355			Hedycaryol	2.01	

Retention times	Dried small ginger (area/high)		Dried big ginger (area/high)		Dried red ginger (area/high)	
12.396					Unknown	4.82
12.509					Isospatulenol	1.35
12.589					1H-Benzo-cycloheten-7-ol, 2,3,4,4a,5,6,7,8-octahydro-1,1,4a,7-tetramethyl	1.42
12.650					4-Bromo-1-naphthalen-amine	1.41
12.786					Beta-maaliene	1.29
12.954					3-Oxatricyclo [20.8.0.0E7,16] triaconta-1(22), 7(16),9,13,23,29′	2.22
13.207					Beta-guaiene	1.29
13.318					6-Isopropenyl-4,8A-dimethyl-3,5,6,7,8,8A-hexahydro-1H	1.41
13.473					Zierone	1.15
14.774					9-Octadecenoic acid, methylester	

Table 1.
Component profile of volatile oil of small, big and red ginger.

Surprisingly, the data of gas-chromatography-mass spectrophotometry (GC-MS) showed that the percentage of zingiberene and sesquiphellandrene, which are mentioned in the WHO Monograph as responsible substances for specific flavour of ginger, is only 1.27 and 1.22%. The biggest component is z-citral or neral that represents 6.04% in small ginger and 6.49% in big ginger. As mentioned above, the volatile oil of red ginger in point view of its components showed to be totally different from small and big ginger. This is in line with the odour of the rhizome. The odour of small and big ginger is stronger than that of red ginger but the pungent taste of red ginger is stronger than small and big ginger (**Figure** 7).

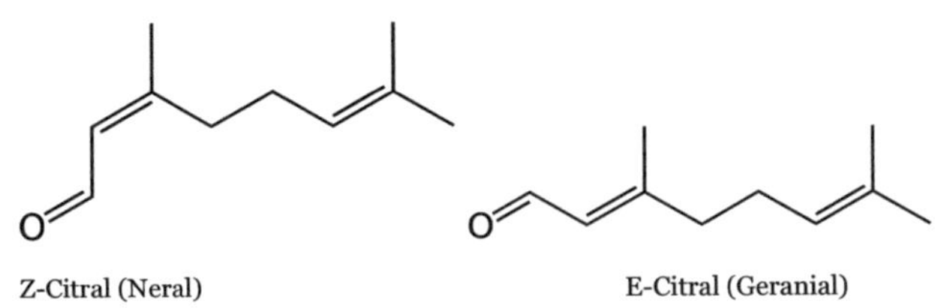

Unfortunately our bibliography does not have data of the biggest substance of red ginger. The fragmentation according to mass spectrophotometry is shown in **Figure 8**.

The profile of fragmentations until m/e 190 is similar to other constituents but between 190 and 220 is rather difficult to calculate. It is not methyl, acetyl, ethyl or other alkyl. It is interesting to be programmed in the future research.

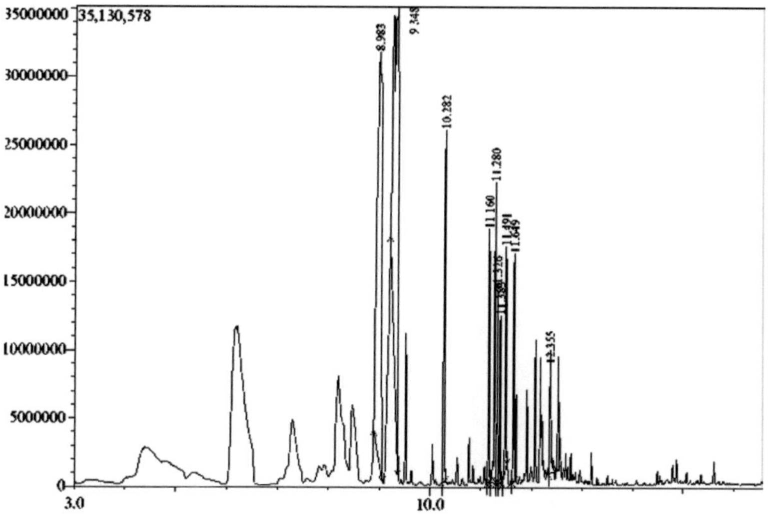

Figure 6.
Profile of gas chromatography of small ginger.

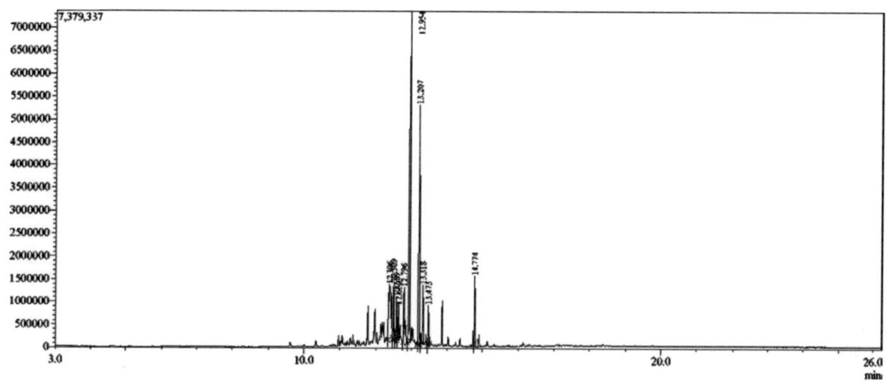

Figure 7.
Profile of gas chromatography of red ginger.

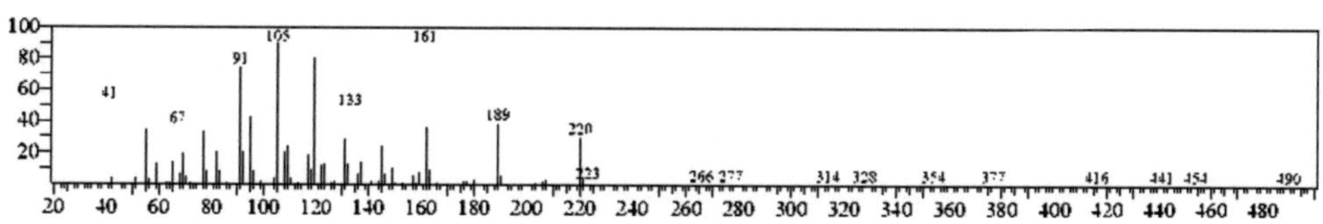

Figure 8.
Fragmentation recorded of unknown peak of gas chromatography profile of red ginger.

Acknowledgements

All data mentioned in figures or table are produced individually in the framework of research of the author (Suwijiyo Pramono) and were performed at Faculty of Pharmacy, Universitas Gadjah Mada, Yogyakarta, Indonesia.

Author details

Suwijiyo Pramono
Department of Pharmaceutical Biology, Faculty of Pharmacy, Universitas Gadjah Mada, Yogyakarta, Indonesia

*Address all correspondence to: suwijiyo_pramono@yahoo.com

References

[1] Ministry of Health Republic Indonesia. Indonesian Herbal Pharmacopeia. 1st ed. Jakarta, Indonesia: Directorate General Pharmaceutical Services and Medical Devices, Republic Indonesia; 2012

[2] Mardisiswoyo S, Rajakmangunsudarso H. Cabe Puyang Warisan Nenek Moyang. Jakarta, Indonesia: Balai Pustaka; 1987

[3] Ministry of Health Republic Indonesia. Vademekum of Medicinal Plants for Jamu Saintification. Vol. 1. Revised ed. Jakarta, Indonesia; 2012

[4] de Guzman CC, Siemonsma JS, editors. Plant Resources of South-East Asia, No.13 Spices. PROSEA; Bogor, Indonesia; 1999

[5] Available from: https://www.google.com/search?client=firefox-b-d&q=preserved+ginger

[6] Pramono S. JAMU, in Indonesian Daily Life and Industry. Toyama, Japan: Institute of Natural Medicine, University of Toyama; 2007

[7] Thomson N, Al-Qattan KK, Al-Sawan SM, Alnaqeeb KI, Ali M. The use of ginger (*Zingiber officinale* Rosc.) as potential anti-inflammatory and antithrombotic agent. Prostaglandins, Leukotrienes, and Essential Fatty Acids. 2002;**67**:475-478

[8] Ghazanfar SA. Handbook of Arabian Medicinal Plants. Boca Raton: CRC Press; 1994

[9] Bisset N, editor. Herbal Drug and Phytopharmaceutical: A Handbook for Practice on Scientific Basis, Medpharm Scientific. Boca Raton: Stuttgart and CRC Press; 1994

[10] Schimid R, Sehick T, Steffen R, Tschopp A, Wilk T. Comparison of seven commonly used agents for prophylaxis of seasickness. Journal of Travel Medicine. 1994;**1**(4):203-206

[11] Philips S, Ruggier R, Hutchinson SE. *Zingiber officinale* (ginger)—An antiemetic for day case surgery. Anaesthesia. 1993;**48**(8):715-717

[12] Borreli F, Aviello G, Pittler MH, Izzo AA. Effectiveness and safety of ginger in the treatment of pregnancy induced nausea and vomiting. Obstetrics and Gynecology. 2000;**106**(3):640

[13] WHO. WHO Monographs on Selected Medicinal Plants. Vol. 1.1999. pp. 277-287

[14] Bensky D, Gamble A, editors. Chinese Herbal Medicine: Materia Medica. Seattle, Washington: Eastland Press, Incorporated; 1993

[15] Sepahvand R, Esmaeili-Mahani S, Arzi A, Rasoulian B, Abbasnejad M. Ginger (*Zingiber officinale* roscoe) elicits antinociceptive properties and potentiates morphin-induced analgesia in rat radiant heat tail-flick test. Journal of Medicinal Food. 2010;**13**(6):1347-1354

[16] Jia Y, Zhao J, Zhang L, Sun B, Bao Mli F, Shen J, et al. Analgesic and antiinflammatory effects of ginger oil. Chinese Herbal Medicines. 2011;**3**(2):150-155

[17] Raji Y, Udoh US, Oluwadara OO, Akinsomisoye OS, Awobajo O, Adhesoga K. Anti-inflammatory and analgesic properties of the rhizome extract of *Zingiber officinale*. African Journal of Biomedical Research. 2002;**5**:121-124

[18] Gunathilake K, Rupasinghe V. Recent perspectives on the medicinal potential of ginger. Archived Journals:

Botanics, Target and Therapy. 2015;**5**:55-63

[19] Yoshikawa M et al. Qualitative and quantitative analysis of bioactive principles in Zingiberis rhizoma by means of high performance liquid chromatography and gas liquid chromatography. Yakugaku Zasshi. 1993;**113**:307-515

[20] Kawakishi S, Morimitsu Y, Osawa T. Chemistry of ginger components and inhibitory factors of arachidonic acid cascade. In: Ho C-T et al., editors. Food Phytochemicals for Cancer Prevention II. Washington, DC: American Chemical Society; 1994

[21] Schwertner HA, Rios DC. High-performance liquid chromatographic analysis of 6-gingerol, 8-gingerol, 10-gingerol, and 6-shogaol in ginger-containing dietary supplements, spices, teas, and beverages. Journal of Chromatography. 2007;**856**(1-2):41-47

[22] Cafino EJV, Lirazan MB, Marfori EC. A simple HPLC method for analysis 6-gingerol in multiple shoot cultures of ginger. International Journal of Pharmacognosy and Phytochemistry Research. 2016;**8**(1):38-42

Section 2

Cultivation and Improvement

Chapter 3

Biotechnology and Crop Improvement of Ginger (*Zingiber officinale* Rosc.)

Neeta Shivakumar

Abstract

Ginger is the third most important spice used for its medicinal properties in day to day life. Ginger is one of the widely studied plants for its biochemical and medicinal properties. Biotechnological tools have played a pivotal role in the improvement of this plant species. Many in vitro techniques namely micropropagation techniques, somatic embryogenesis, somatic hybridization, germplasm conservation, transgenics and mutation breeding have been widely studied whereas less studied for haploid production, and cryogenic in ginger. Many of these have been used in the recent times for the improvement of ginger mainly because of the vegetative mode of propagation. Most varietal improvement programs of this species are confined to evaluation and selection of naturally occurring clonal variations. Problems faced in ginger breeding have so far been the very low genetic variation in ginger plant. Wide genetic variation is needed in plant breeding in order to search ideal plant types during the process of selection. Although traditional mutation breeding has lost its preeminent position, induced mutations continue to be in great demand with the assistance of various biotechnological tools. In vitro culture techniques provide an alternative means of plant propagation and a tool for varietal improvement. Here, is an attempt made to collect the information on the studies made in this regard and present the current status of research in ginger.

Keywords: ginger, *Zingiber officinale*, *in vitro* culture, biotechnological tools, crop improvement

1. Introduction

Ginger (*Zingiber officinale* Rosc.) belongs to the family Zingiberaceae, is an important tropical horticultural plant and an important spice crop used in various medicinal and culinary preparations. Ginger consumption is known for its health benefits and widely known to be used in Ayurvedic formulations and Chinese medicine. It is rich in secondary metabolites namely the oleoresins and shogaols contributing widely the pungency and flavors [1]. It is stimulative in nature and helps in relieving the indigestion, stomach ache, diarrhea and nausea. It is widely used as to cure common cold, cough and congestion. The clinical studies have demonstrated it to be anti-emetic, anti-ulcer, anti-platelet, anti-inflammatory, and antioxidant in nature. Ginger has many uses in the home remedies and can be used to help arthritis, diarrhea, flu, headache, heart and menstrual problems, diabetes, stomach upset and motion sickness [2]. Wide studies have been taken up involving ginger to cure complex diseases such as cancers to the chronic conditions of migraines.

Ginger has been widely used in various medicines viz., Ayurveda, Unani and Chinese medicines to cure many of the health problems. It has been used in various ways either in the food directly as or as a part of the formulations in medicines to cure many of the ailments such as sore throat, muscle strains, chronic cough, asthma, headaches, diabetes, relief of nausea and flatulence. Studies have proved ginger to have anti-inflammatory effects, anti-cancerous specially colorectal and ovarian and antiemetic anti-platelet activity [3].

Ginger is carminative, pungent, stimulant, used widely for indigestion, stomach ache, malaria and fever. It is chiefly used to cure diseases due to morbidity of Kapha and Vata. Ginger with lime juice and rock salt increases appetite and stimulates the secretion of gastric juices. It is said to be used for abdominal pain, anorexia, arthritis, atonic dyspepsia, bleeding, cancer, chest congestion, chicken pox, cholera, chronic bronchitis, cold extremities, colic, colitis, common cold, cough, cystic fibrosis, diarrhea, difficulty in breathing, dropsy, fever, flatulent, indigestion, disorders of gall bladder, hyperacidity, hypercholesterolemia, hyperglycemia, indigestion, morning sickness, nausea, rheumatism, sore throat, throat ache, stomach ache and vomiting. Ginger forms an important constituent of many pharmacopoeial Ayurveda formulations [4].

1.1 Botany of the ginger plant

The ginger plant is an erect herbaceous perennial growing from one to three feet in height. The stem is surrounded by the sheathing bases of the two-ranked leaves. The plant is erect, aerial shoots (pseudostem) with leaves, and the underground stem (rhizome). The fleshy and fibrous roots of ginger have indefinite growth from the base of the sprouts. These are the fibrous roots, and the number of such roots keeps on increasing with the growth of tillers [5] (**Figure 1**).

A club-like spike of yellowish, purple-lipped flowers have showy greenish yellow bracts beneath. Unfortunately, ginger rarely flowers in cultivation [5] (**Figure 2**).

Ginger is the modified rhizome stem having nodes with scale leaves and internodes. Except for the first few nodes, all the nodes have axillary buds. Generally, the seed rhizome bit usually termed as setts is used for planting. There may be one or more apical buds on it. Normally only one bud becomes active. When large pieces are

Figure 1.
Ginger cultivation. Source: https://in.images.search.yahoo.com/.

Figure 2.
Flowering in ginger. Source: https://in.images.search.yahoo.com.

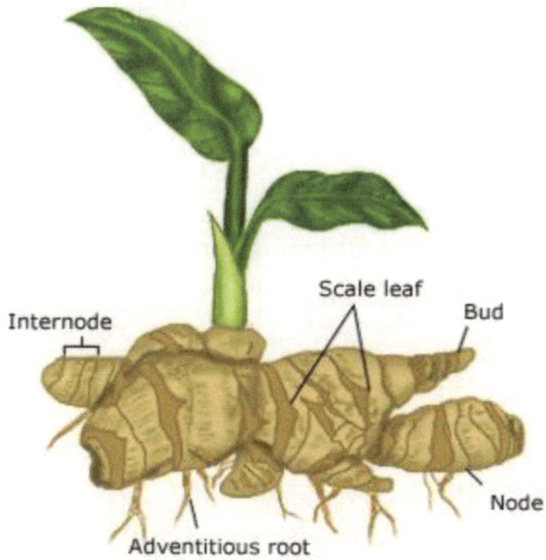

Figure 3.
Botany of ginger plant. Source: https://in.images.search.yahoo.com/.

used, more than one bud may develop simultaneously. If more than one branch from the parent rhizome is responsible for the ultimate growth and development of the adult rhizome, the branches of the mature rhizome lie in the same plane (**Figure 3**).

1.2 Color and appearance

Ginger can be found in Yellowish brown or light brown colors. Scraped rhizome with buff external surface showing longitudinal striations and occasional loose fibers, outer surface dark brown [1] and more or less covered with cork which shows conspicuous, narrow, longitudinal and transverse ridges (**Figure 4**).

1.3 Nutrient composition

Nutrient composition of ginger includes 80% moisture, 2.3%protein, 0.9%fat, 1.2% minerals, 2.4% fiber and 12.3% carbohydrates. The minerals present in ginger are iron, calcium and phosphorous. It also contains vitamins such as thiamine, riboflavin,

Figure 4.
Modified rhizome stem. Source: https://in.images.search.yahoo.com.

niacin and vitamin C. The composition varies with the type variety, agronomic conditions, curing methods, drying and storage conditions. The rhizome, which is valued for its flavor, contains two constituents such as the essential oils and oleoresins [5].

1.4 Ginger varieties and its constituents

1.4.1 Varieties of ginger

Zingiberus family includes about 50 genera and 1300 species of ginger are known to exit worldwide. They occur in different parts of the world namely Japan, Australia, Haiti, Bangladesh, Jamaica, Sri Lanka, Nigeria. Several cultivars of ginger grown in different ginger growing areas in India and they are generally named after the localities where they are grown. Some of the prominent indigenous cultivars of ginger grown in India are Himachal, Maran, Kuruppampadi, Wayanad, Varada, etc. Exotic cultivars such as Rio-de Janeiro have also become very popular among cultivars. Maran, Nadia, Karakkal, Rigodi are suited for high dry ginger. Varieties like Ernad Chernad, China and Rio-de Janeiro provide high oleoresin content. Sleeve local, Narasapattam, Himachal are suited for high volatile oil. Rio de Janeiro, China, Wayanad, Maran, Varada are suited for green ginger [6].

2. Biotechnology of ginger

Biotechnological tools have played a pivotal role in the improvement of plant species. There are many techniques specially the micropropagation techniques, production of enhanced secondary metabolite production using cell suspension cultures [7]. *In vitro* production of haploids, somatic embryogenesis, somatic hybridization [8], germplasm conservation and storage, recombinant DNA technology and transgenic [9]. Many of these have been used in the recent times for the improvement of ginger mainly because of the vegetative mode of propagation and those techniques have been discussed below.

2.1 Micropropagation of ginger

Ginger is one of the important spice crops in India under the family Zingiberaceae. This plant is a highly important horticultural crop and plays a very important role in pharmaceuticals, food and beverage industries. In a vegetatively propagated plant like ginger, risk of systemic Infections by root knot nematodes, bacterial wilt from the propagules is very high. Heavy losses in ginger productions is due to diseases caused by bacterial wilt (P-50), soft rot (Pythium) and rhizome rot (E-Oxy). Therefore, the diseases are mainly transmitted by rhizome propagules; production of disease free clones is necessary. Micropropagation using tissue culture technique can be the alternative method. *In vitro* culture technique provides an alternative means of plant propagation and a tool for crop improvement [10]. Clonal multiplication of ginger through shoot multiplication also has been reported. Clonal propagation is achieved through direct organogenesis or indirect organogenesis [11]. Establishment of the explants plays a very important role in micropropagation. The explants usually used are the rhizome buds. Leaf explants, internodes and roots. However, the adventitious buds have been widely used. The buds have the preformed primordia for the direct organogenesis. Nutrients are provided for the development of the shoots from these buds. In indirect organogenesis the explants is subjected to enter the callus phase and then dedifferentiate into plantlets. Callus culture is proposed for rapid proliferation of plant cells [12]. However, contamination plays a very important role in the establishment. In order to achieve contamination free cultures the buds thoroughly washed with running water and with the detergents namely Tween 20 under running water. These dust free/soil free buds are later subjected with surface sterilants such as mercuric chloride and ethanol, Followed by washing with distilled water and inoculated. The following are the established cultures seen in the author's lab (**Figure 5**).

These shoots are subjected to multiple shoot formation steps which needs the specific media for multiplication. This requires the lower concentration of auxins and cytokinins with 2 mg/l NAA and 0.1 mg/l BAP was used for the multiplication [7]. Following are the established cultures seen in the author's lab (**Figure 6**).

After multiplication is the elongation step where the gibberllic acid is added. However since this step was not required the shoots are transferred to rooting media (**Figure 7**).

Multiple shoots transferred for rooting with MS medium supplemented with 2 mg/l NAA and 0.1 mg/l BAP. Many other researchers have tried different concentrations growth enhancers. After rooting the shoots have been primarily acclimatized by transferring to the netted pots filled with sterilized peat mixture and kept in the growth chamber by maintaining a humidity of 80% and the light period 16 ± 8 hrs. After 2 weeks transferred to green house for secondary acclimatization of the plants in netted pots [7].

2.2 Enhanced secondary metabolite production

There are various studies conducted in ginger to compare the secondary metabolite production both *in vitro* and *in vivo*. The In vitro culture did not enhance the production of the gingerols and gingiberene. However, the in vivo produced the same amount of secondary metabolites [11]. In another study the influence of mevalonic acid (MVA) increased the production of 6-gingerol in in vitro grown callus cultures compared to other precursors and biotic elicitors. Other active substances did not influence the production of 6-shogaol, 8-gingerol and 10-gingerol.30 μg/100 mg callus FW was recorded with 1 mg/l MVA [12]. Many elicitation studies need to be conducted for different varieties of ginger.

2.3 *In vitro* production of haploids

Studies have not been done on haploid production in ginger. The probable reason may be the poor flower setting which makes androgenic or gynogenic studies nearly impossible.

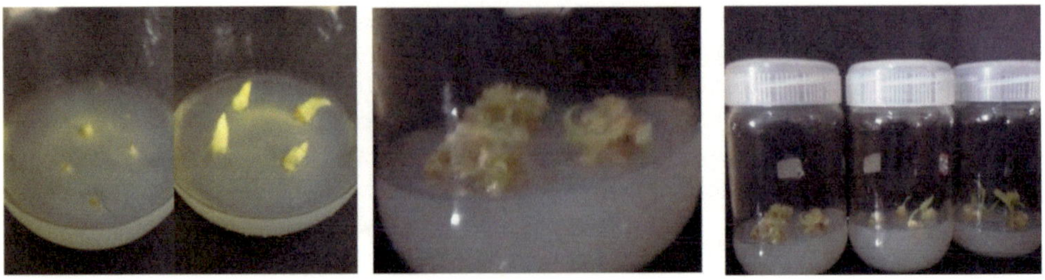

Figure 5.
Stages of micropropagation through callus (source: author's research work).

Figure 6.
Fully grown ginger explants via micropropagation (source: author's research work).

Figure 7.
Stages in hardening of ginger explants (source: author's research).

2.4 Somatic embryogenesis

It has been studied in ginger using leaf explants The embryogenic callus was maintained using MS media with dicamba as the best growth regulator [13]. Somatic embryogenic studies were taken using meristematic explants to produce pathogen free seedling [14]. Not many studies have been taken up with respect to different varieties of ginger for somatic embryogenesis.

2.5 Somatic hybridization

Plays a pivotal role in the improvement of the crop species specially the vegetatively propagated ones. Different methods of somatic hybridization have been proposed namely the physical and chemical method of fusion techniques. Chemical method is commonly used technique using polyethylene glycol [14]. In this study, two different species of ginger ['Lushan Zhangliangjiang' + 'Chenggu Huang Jiang' (LZ + CH)] were fused to obtain a hybrid and later regenerated after 15 months. This technique could be adapted to create many other hybrids were the varieties are low yielding in terms of quantity and quality of rhizomes.

2.6 Germplasm conservation and storage

Germplasm conservation is an important technique for conservation of the plant diversity for any of the country. The plant species can be conserved mainly through in situ and ex situ methods. In situ methods have many disadvantages mainly being the land requirement for cultivation, seasonal changes which may lead to destruction of the plant material and high cost incurred during cultivation. However Ex situ conservation in recent days has gained more prominence and importance utilizes the in vitro techniques for the conservation of the species. In vitro techniques are the widely used techniques for ex situ conservation and utilizes the principle of the reduced amount of nutrients supplied to the explants and in turn would reduce the growth of plant and less attention can be given for a certain amount of time. Studies on the ginger germplasm conservation have shown that the cultures can be maintained on half strength MS with reduced amount of sucrose/carbohydrate source and maintain the cultures at ambient temperature of 22 ± 2oC. The cultures can be maintained for 200–240 days with 75–81 percentage of survival. Half strength MS with 20 g/l of sucrose extended the culture span to 360 days [14]. Usage of polypropylene caps is also reported to maintain the cultures for 7 months. Around 100 accessions have been maintained at Indian Spice Research Institute (IISR) and In vitro conservation provides a promising tool for maintaining the ginger cultivars. However, there are no reports available for the successful cryopreservation strategy.

2.7 Recombinant DNA technology

2.7.1 Transgenics

Nowadays, researches are taking place in bringing out transgenic varieties of ginger. Trasgenics have been developed to overcome crop loss due to disease, lack of resistance, and post-harvest losses. Since ginger is vegetatively propagated, the genetic variability is limited. When compared to other spices, transgenic developments are very few in number in case of ginger.

Transient expression of GUS was successfully induced in ginger embryogenic callus bombarded with plasmid vector pAHC 25 and promoter Ubi-1(maize ubiquitin) callus tissue [15]. Helium bombarded ginger embryo-genic calli with microprojectiles (1.6 μm gold particles) using 'BioRad' PDS-1000/He gene gun at 900 and 1100 psi helium pressure with the target distance of either 6 or 9 cm. The pAHC 25 vector used contained GUS (β-glucuronidase) and BAR (phosphinothricin - acetyl transferase) as reporter and selectable marker genes respectively. The best GUS score was obtained when the target distance was 9 cm with 900 psi helium pressure. The GUS score of 133 blue spots per cm^2 indicated not only the optimization and efficiency of the biolistic process, but also the ability of the ubiquitin promoter to drive the expression of the reporter gene. [16].

It was reported that the genetic transformation of ginger buds through somatic embryogenesis. They found that young buds had very good embryogenic potential and were superior to other explants. The transformation protocol included 3 day pre-culture of explants on callus induction medium, bacterial (*Agrobacterium tumefaciens* strainEHA105/p35SG) dilution of 1:20 (v/v) as the initial inoculum, an infection time of 5 minutes, 2 day co-cultivation with Agrobacterium and post cultivation on callus induction medium with 100 mgL^{-1} kanamycin and 300 mg L^{-1} cefotaxime under darkness for 2 weeks, followed by a 16/8 h photoperiod regime. Acetosyringone was effective at a concentration of 200 μm for vir induction. With young bud as explant, a transformation frequency of 1.1 to 2.2% was noticed. The callus growth was very slow in the presence of antibiotics [17].

3. Breeding of ginger

Breeding of ginger is seriously handicapped by poor flowering and seed set. Most crop improvement programs of this species are confined to evaluation and selection of naturally occurring clonal variations. Problems faced in ginger breeding have so far been the very low genetic variation in ginger plant. Wide genetic variation is needed in plant breeding in order to search ideal plant types during the process of selection [18]. Although traditional mutation breeding has lost its pre-eminent position, induced mutations continue to be in great demand with various biotechnological tools. The methods of mutation induction and analyses of mutants have witnessed great changes in recent years. *In vitro* culture techniques provide an alternative means of plant propagation and a tool for crop improvement (Vasil, 1988). The advent of tissue culture technology enables small and easily handled amount of tissue to be treated. Excised stem tips or callus growing on standard nutrient medium could be treated and the explants are grown to maturity and evaluated for useful mutations. Advanced *in vitro* manipulations such as somatic embryogenesis and single cell cultures also reduce the problem of chimerism often encountered in the induction of mutations on vegetatively propagated plants (Chopra, 2005).

3.1 Mutational breeding in ginger

Mutational breeding has a very prominent role to play in breeding a new variety specially the ones which are clonally propagated. The traditional methods of mutation breeding involves the use of the mutagens in creating the mutation and check for the mutated genes to pass from one generation to the other. These trails consume lot of time, laborious and needs the involvement of large population. With the advent of Biotechnology techniques, the process is made easier in mutation breeding wherein the explants are treated with the mutagens and subjected to screening under *in vitro* conditions. The other method of creating variation is through somaclonal variation, somatic hybridization induced under in vitro conditions and recently is through transgenic technology.

Mutation breeding can be taken up by treating large number of rhizomes under in vivo conditions and screen for the phenotypic and genotypic characters. These studies require large number of planting material, cost and space. The methods of mutation induction and analysis of mutants have witnessed a great change in recent years. With the advances in recent biotechnological studies, mutation studies can be taken under in vitro conditions. The explants namely the adventitious buds, leaf explants, stem and roots and generate the plants through direct regeneration or indirect regeneration.

Ginger is severely affected bacterial wilt and rhizome rot diseases. The lack of genetic variability among the genotypes for disease resistance is one of the bottlenecks in ginger genetic improvement. Studies have been done to induce mutability and radio sensitivity of the ginger genotypes using different doses gamma rays.

Figure 8.
Stereo microscope images of the mutants.

Different mutation frequencies and width of mutation spectra were induced under the action of different concentrations of the gamma rays (0.5–1.2kR). High frequency of chlorophyll mutants (5.13%) indicates mutability of ginger. The spectrums of chlorophyll mutations (albino, xantha, and chlorina) were observed and grouped. The overall mutation spectrum for ginger showed that xantha occurred with the highest frequency, followed by chlorina and albino. The mutagenic effectiveness decreased with the increase in dose of mutagen that negative relationship between effectiveness and dose of mutagen. Compared to control, wide variability was recorded for various morphological characters under different doses of gamma rays. Artificial screening of mutants against Ralstonia solanacearum and Pythium sp. led to identification of six potential mutants against these two pathogens [18].

The chlorophyll mutation frequency in mutation generation is the most dependable index for evaluating the genetic effects of mutagenic treatments. The effect of Ethyl Methane Sulfonate (EMS), Sodium Azide and Colchicine on induction of different types of chlorophyll mutants have been widely used to determine their frequency in ginger. There are many mutagens which can be used namely physical viz., UV rays, X rays, Gama rays mutagens etc., chemical EMS, Sodium azide and Choline biological [19] (**Figure 8**).

Ethyl Methane Sulfonate (EMS) is a mutagenic which is carcinogenic in nature. It produces random mutations in the genetic material. The EMS concentrations used varied from 5 mM to 10 mM. Following are the chlorophyll variations observed and classified as follows:

- Xantha: Leaves with little or no chlorophyll but have carotenoid pigmentation and are yellow.
- Striata: They have yellow or white longitudinal bands alternating with green color.
- Maculata: Have spots where chlorophyll and/or carotene have been destroyed.

The frequency of mutation can be calculated using the below formula:

$$\text{Mutation frequency (\%)} = \frac{\text{No. of mutants}}{\text{Total no. of plants scored}} \times 100 \tag{1}$$

Mutation studies have been taken up in the indigenous varieties of ginger from India namely Bidar, Himachal and Humnabad. They observed the variation in 3 months plants of the mutant varieties and were studied with respect to various growth parameters like survival percentage, sprouting percentage, plant height, number of tillers, etc. They found that the treatment with chemical mutagens had significant effect on all the growth and yield parameters. Chlorophyll mutations spectrums were observed and grouped. Mutation frequency was calculated for VM_0 and VM_1 generation, mainly with the occurrence of maculata highest, followed by xantha and striata. More number of mutants was found with the plants treated with EMS, followed by colchicines and sodium azide. The albinism was rarely exhibited by plants [20].

Similarly, Sri Lankan varieties have been studied for using EMS for the yield and quality parameters studied different chlorophyll variants induced with gamma rays and EMS and combination of both the treatments. Physical mutagens are mainly the gamma radiations and UV radiations which have been studied. Most effective being the Gamma radiations and can be used widely not only in ginger but also other crops [20].

3.2 Role of markers in ginger breeding

Ginger (*Zingiber officinale*) is a crop grown widely in tropical and subtropical regions which is of medicinal, economic and horticultural importance. There are more than 150 varieties of ginger. In species like ginger which reproduces mainly through vegetative modes, the chances of crop improvement are very limited. A molecular marker is a molecule contained within a sample taken from an organism or other matter. It can be used to reveal certain characteristics about the respective source. Markers have played a tremendous role in breeding and conservation of species like ginger [21].

Randomly Amplified Polymorphic DNA (RAPD) Markers are used for plant species like ginger because no specific information about the genome is available. Ginger genome is one among the poorly studied genomes and information available is very little. Using phylogenetic analysis and metabolic profiling of diversity within and among *Zingiber* species was investigated. It was found that *Zingiber* from different geographical locations were indistinguishable [22]. Clones, cultivars, varieties, accessions or genotypes can be distinguished by RAPD markers with high resolving power value. RAPD as a marker is well-established in generating reproducible polymorphic bands. Variation among the induced mutant clones could be used to support the selection process at the early stage of the plant. RAPD markers can differentiate ginger populations based on their collection sources. Populations of *Z. zedoaria* and *Z. officinale* were distributed into two groups that exist in hill areas and plain areas, respectively, while farm collections showed wide genetic diversity. RAPD has proven to be an effective tool for detecting genetic diversity at the interspecific and intraspecific levels [23, 24]. Thus, it has been proves that RAPD markers can be used for identification and classification of ginger species.

Amplified fragment length polymorphism (AFLP) markers are capable of disclosing a greater number of polymorphisms with a single reaction. Thus, this marker has become the main tool in genetic marker technologies [25]. The genetic relationship within a species or genus could be determined using AFLP markers. The polymorphic bands were quite low within species. Using AFLP analysis it was found that ginger phylogenetic relationships mostly corresponded to their morphological characteristics and modes of reproduction. Genetic diversity patterns within different genome sizes of Curcuma populations were confirmed to be influenced by the mode of reproduction. ALFP markers could provide species-specific identification for examined species and produce a large number of reproducible markers to assess diversity across the nuclear genome. Three species of *Zingiber*, *Z. officinale, Z. montanum* and *Z. zerumbet*, were identified by using this DNA fingerprinting marker. This shows that ALFP can be used as a reliable parameter for identifying species [26].

Simple sequence repeat (SSR) marker is believed to be one of the most powerful markers in studying genetic diversity. It can accurately assess the level of genetic diversity within a germplasm of any crop. Available polymorphic microsatellite markers have been found in the ginger species *Zingiber officinale* and *Curcuma longa* [27]. Until now 56 genomic SSR markers and 17 EST-SSR markers have been developed for *Curcuma longa*, while only eight genomic SSR markers have been reported in *Z. officinale*. SSR markers displayed better polymorphism results compared to ISSR and IRAP markers [28]. The development and characterization of microsatellite markers for ginger species would be useful for future studies evaluating genetic diversity and genetic divergence among species. SSR markers could be used for ginger breeding improvement programs.

Inter-Simple sequence repeats (ISSRs) are semi-arbitrary markers amplified by PCR in the presence of one primer corresponding to a target microsatellite [25]. They are randomly distributed throughout the genome. The main advantage of ISSRs is that the amplification does not require genome sequence information and produces high fragments. ISSR is effective for differentiating relationships among closely related ginger cultivars [29].

This technique has been confirmed to show higher polymorphisms compared to RAPD and AFLP markers when tested on Curcuma species [30]. ISSR markers were found to be less informative compared to AFLP in the case of wild the ginger species, *Z. moran* and cultivars of Northwest Himalayan [28]. The differences in resolving power of these markers is due to the difference portions of the genome that are targeted. Thus, it is essential to apply different molecular techniques in diversity studies of any crops.

Combination of RAPD and ISSR is also used in identification of ginger species. Several ginger studies applied both RAPD and ISSR as molecular markers. RAPD and ISSR have been used for genetic diversity studies in various plant species including ginger. RAPD markers are reported to be more suitable for genetic diversity analysis. ISSR markers are more reproducible compared to RAPD markers. Therefore, the percentage of polymorphisms using ISSR markers was quite higher than RAPD markers. Combining data from the two markers would give a better result in terms of species differentiation. RAPD and ISSR have been used to assess genetic diversity among micro-propagated and cloned ginger species [31]. The aim of the studies was to reveal monomorphic bands to confirm genetic stability or uniformity.

Other markers that have been used in ginger species include restriction fragment length polymorphism (RFLP), sequence characterized amplified region (SCAR), nucleotide binding site (NBS) and single nucleotide polymorphism (SNP) [31] have been used to identify and characterize ginger species.

Author details

Neeta Shivakumar
Department of Biotechnology, R V College of Engineering, Bangalore, India

*Address all correspondence to: neeta@rvce.edu.in

References

[1] Bhagyalakshmi B, Narasimhan S, Narendra S. The yield and quality of ginger produced by micropropagated plants as compared with conventionally propagated plants. Journal of Horticultural Science. 1993;**4**:645-651

[2] O'Hara M et al. A review of 12 commonly used medicinal herbs. Archives of Family Medicine. 1998;**8**(5):376

[3] Mowrey DB et al. Motion sickness, ginger, and psychophysics. Clinical Trials. 1982;**1**(8273):655-657

[4] Mustafa T, Srivastava KC. Ginger (*Zingiber officinale*) in migraine headache. Journal of Ethnopharmacology. 1990;**29**:267-273

[5] Bhatt N, Waly MI, Essa MM, Ali A. Ginger: A Functional Herb. Nova Science Publishers; 2013. pp. 51-71

[6] Bag BB. Ginger processing in India (*Zingiber officinale*): A review. International Journal of Current Microbiology and Applied Sciences. 2018;7(4):1639-1165

[7] shivakumar N, Agrawal P. Callus induction and regeneration from adventitious buds of *Zingiber officinale*. Asian Journal of Microbiology, Biotechnology & Environmental Sciences. 2014;**16**(4):881-885

[8] Kackar A, Bhat SR, Chandel KPS, et al. Plant regeneration via somatic embryogenesis in ginger. Plant Cell Tissue Culture. 1993;**32**:289

[9] Baskaran P, Jayabalan N. An efficient micropropagation system for Eclipta, alba—A valuable medicinal herb. In Vitro Cellular & Developmental Biology – Plant. 2005;**41**:532-539

[10] Nirmal Babu K et al. Protocols for in vitro propagation, conservation, synthetic seed production, embryo rescue, microrhizhome production, molecular profiling, and genetic transformation in ginger. Methods in Molecular Biology. 2016;**1391**:403-426

[11] Seran TH. In vitro propagation of ginger (*Zingiber officinale*) through direct organogenesis: A review. Pakistan Journal of Biological Sciences. 2013;**16**(24):1826-1835

[12] El-Nabarawya MA, El-Kafafia SH, Hamzab MA, Omara MA. The effect of some factors on stimulating the growth and production of active substances in *Zingiber officinale* callus cultures. Annals of Agricultural Science. 2015;**60**(1):1-9

[13] Faria RT, lllg RD. Micropropagation of *Zingiber spectbile* Griff. Scientia Horticulturae. 1995;**62**:135-137

[14] Rostiana O, Syahid SF. Somatic embryogenesis, from meristem explants of ginger. Biotropia. 2008;**15**(1):12-16

[15] NirmalBabu K, Yamuna G, Praveen K, Minoo D. Cryopreservation of spices genetic resources. INTECH. 2012;**16**:457-484

[16] Abbas MS, Gaber E s I. In vitro propagation of ginger (Zingiber officinale Rosco). Journal, Genetic Engineering & Biotechnology. 2011;**9**:165-172

[17] Babu KN, Minoo D, Geetha SP, Samsudeen K, Rema J, Ravindran PN, et al. Plant biotechnology-its role in improvement of spices. Indian Journal of Agricultural Sciences. 1998;**68**(8):533-547

[18] NirmalBabu K, Suraby EJ, Cissin J, Minoo D, Pradeepkumar T, Parthasarathy VA, et al. Status of transgenics in Indian spices. Journal of Tropical Agriculture. 2013;**51**(1-2):1-14

[19] Suma B, Keshavachandran R, Nybe EV. Agrobacterium tumefaciens mediated transformation and regeneration of ginger (Zingiber officinale Rosc). Journal of Tropical Agriculture. 2008;**46**:38-44

[20] Prasanth D, Suseela Bhai R, Nair R. Induction of Variability in Ginger through Induced Mutation for Disease Resistance, Conference: National Symposium on Spices and Aromatic Crops. 2015. pp. 16-18

[21] Prasannakumar P, Subramanian S, Suresh J, Kannan Bapu JR, Gnanam R. Effect of gamma rays on induction of chlorophyll mutants in ginger genotypes. International Journal of Agricultural. 2016;**6**:105-110

[22] shivakumar N, Agrawal P. The effect of chemical mutagens upon morphological characters of ginger in M0 generation. Asian Journal of Microbiology, Biotechnology & Environmental Sciences. 2018;**20**:126-135

[23] Ravinderan PN, Nirmal BK, Shiva KN. Botany and crop improvement of ginger. In: Ravinderan PN, Nirmal BK, editors. Ginger: The Genus Zingiber. New York: CRC Press; 2005. pp. 15-85

[24] Nayak S, Naik PK, Acharya L, Mukhergee AK, Panda PC, Das P. Assessment of genetic diversity among 16 promising cultivars of ginger using cytological and molecular markers. Zeitschrift für Naturforschung. 2005;**60c**:485-492

[25] Das A, Kesari V, Satyanarayana VM, et al. Genetic relationship of curcuma species from Northeast India using PCR-based markers. Molecular Biotechnology. 2011;**49**:65-76

[26] Donipati P, Sreeramulu SH. Relationships among six medicinal species of curcuma assessed by RAPD markers. International Journal of Recent Scientific Research. 2015;**6**(8):5909-5912

[27] Kumar P, Gupta VK, Misra AK, et al. Potential of molecular markers in plant biotechnology. Plant Omics. 2009;**2**(4):141-162

[28] Kaewsri W, Paisooksantivatana Y, Veesommai U, Eiadthong W, Vadrodaya S. Phylogenetic analysis of Thai Amomum (Alpinioideae: Zingiberaceae) using AFLP markers. Kasetsart Journal (Natural Science). 2007;**41**:213-222

[29] Sigrist MS, Pinheiro JB, Azevedo-Filho JA, et al. Development and characterization of microsatellite markers for turmeric (Curcuma longa). Plant Breeding. 2010;**129**:570-573

[30] Pandotra P, Gupta AP, Husain MK, et al. Evaluation of genetic diversity and chemical profile of ginger cultivars in the North-Western Himalayas. Biochemical Systematics and Ecology. 2013;**48**:281-287

[31] Jatoi SA, Kikuchi A, Yi SS, et al. Use of rice SSR markers as RAPD markers for genetic diversity analysis in *Zingiberaceae*. Breeding Science. 2006;**56**:107-111

Chapter 4

Cultivation of Ginger in Sikkim under an Organic System

Vijayan A.K., B.A. Gudade, Ashutosh Gautam, T.N. Deka, S.S. Bora, K. Dhanapal and A.B. Remashree

Abstract

Ginger is grown extensively throughout India due to its high value and ginger is used for wide range of purposes like in confectionery, traditional medicine for stomach ache, food additives and pickles. The major ginger-producing states include Kerala, Assam, Gujarat, Orissa, Sikkim, Meghalaya, Arunachal Pradesh and Mizoram. It is one of the main cash crops in Himalayan state of Sikkim. In Northeast India, especially in Sikkim, ginger serves as a source of income for small and marginal farmers. It is cultivated in a varying degree of altitude, but the elevation of 1500 above msl is found to be more suitable. Ginger is a tropical plant, and warm, humid climate is the most ideal for ginger cultivation; it grows best in rich soil and shady places. Sikkim has its own indigenous cultivars of ginger, and the prominent varieties that are being cultivated in Sikkim are Bhaise, Gorubathane, Majhaule, Tange, Patle and Jorethang. November to January after 8–9 months of sowing is the optimum time for harvesting ginger; however, this follows the market demand dynamics in Sikkim. Under organic conditions, farmers normally get a yield of 90–100 q/ha depending on ginger cultivation practices. Progressive farmers by adopting improved method of ginger cultivation get on an average of Rs. 150,000 per hectare (benefit-cost ratio varied from 3.50 to 3.80).

Keywords: *Zingiber officinale* Rosc, Sikkim, organic system

1. Introduction

Ginger botanically known as *Zingiber officinale* Rosc. is a crop of family Zingiberaceae and grown extensively throughout India. It is a perennial herbaceous flowering plant which grows annual pseudo-stems whose rhizome, ginger root or ginger is widely used as a spice and a folk medicine. In India, ginger is used for wide ranges of purposes like confectionery, traditional medicine for stomach ache, food additives, pickles, etc. Traditionally, the rhizome is gathered when the stalk withers; it is immediately scalded or washed and scraped, to kill it and prevent sprouting. In China it is used as a typical traditional medicine that has been widely used for the treatment of indigestion, sore throats, rheumatism and hypertension [1–4].

Ginger has been demonstrated to have various pharmacological activities, such as antiemetic, antiulcer, anti-inflammatory, antioxidant, antiplatelet and anticancer activities [5–8].

The major ginger-producing states include Kerala, Assam, Gujarat, Orissa, Sikkim, Meghalaya, Arunachal Pradesh and Mizoram. Ginger is grown as an intercrop in coconut and areca nut plantations in the states of Kerala, Meghalaya, Orissa and West Bengal and to some extent in Karnataka as well as pure crop in the states of Andhra Pradesh and Tamil Nadu [9, 10].

It is one of the main cash crops in Himalayan state of Sikkim and locally known as "Adua", and its cultivation in the state can be traced in ancestral farming in ancient time. Limboos, Phedamgmas, Rai and Bijuwas used ginger in religious practices as well as medicinal application for different purposes. In Northeast India especially in Sikkim, ginger serves as a source of income for small and marginal farmers. It is cultivated in a varying degree of altitude, but an elevation of 1500 above msl is found to be more suitable. The areas which cover major area under ginger cultivation in the state are Tharpu, Malbansey, Gyalshing, Mangalbaria belt, Chakung, Chumbung and Zoom in the West district; Rhenock, Rongli, Rorathang, Pakyong, Pachekhani, Pandam, Khamdong, Sirwani, Sang and Namziism in the East district; Turuk, Sumguk, Ratepani, Namthang, Melli Dara, Maniram, Namchi, Temi Tarku and Bermiok in the South district; and small pockets of Dzongu and Mangan area of the North district.

2. Crop botany

Ginger is an herbaceous perennial crop belonging to the family Zingiberaceae. It is cultivated for its rhizome which is a modified system. It is an aromatic herb with an underground rhizome and an erect stem, up to 75 cm in height. The leaves are green, simple, alternate, sheathing at the base, sessile, acuminating at the apex, glabrous, 15–20 cm long, narrow and linear-lanceolate with prominent midrib and parallel venation. Inflorescence is a spike on a distinct scape; flowers are densely arranged, bisexual and irregular, each subtended by a persistent scarious bract.

Calyx tubular shortly 3-lobed; corolla bilabiate, tubular below, yellow with purplish spots stamens three in one whorl, one of which is perfect and the other two united to form a labellum; ovary of three carpels, syncarpous, 3-celled, inferior; ovules many on axile placentas; style filiform, lying in a groove of the anther; stigma subglobose, filament of perfect stamen short, anther cells contiguous, connective produced into a beak. Flowering in Sikkim is a common phenomenon, but it flowers only in a specific climatic conditions. Ginger produces clusters of white and pink flower buds that bloom into yellow flowers. It is a perennial reed-like plant with annual leafy stems, about a metre (3–4 feet) tall.

3. Uses

Ginger has a wide range of traditional and present-day uses in Indian households, and Sikkim is not an exception too. In Sikkim conditions, ginger is marketed as fresh and used mostly in salad, tea blends, sauces and additives as well as a sole constituent of curry and pickles, confectionery, curry powder, ginger chocolates and candies and for beverage purposes like ginger cordial, ginger cocktail, ginger tonic, ginger brandy, ginger wine, ginger beer, etc. It has many medical values and is used as carminative and stimulant for gastritis, dyspepsia and flatulence and a remedy for bee bites, etc. Ginger contains a variety of oleoresin and volatile oils which are low-volume high-cost products for marketing and fetches high monetary returns.

4. Climate and varieties

Ginger is a tropical plant, and warm, humid climate is the most ideal for ginger cultivation; it grows best in rich soil and shady places. Ginger can be grown both under rainfed and irrigated conditions. The well-distributed rain during the growing season, i.e. from April to October, is ideal for its cultivation. This is the reason that this crop is grown from 300 to 1500 m above msl in Sikkim, where the temperature remains higher during the growing period. It is partially a shade-loving crop and can be grown in semi-shaded condition. Sikkim has its own indigenous cultivars of ginger, and the prominent varieties which are being cultivated in Sikkim are Bhaise, Gorubathane, Majhaule, Tange, Patle and Jorethang.

5. Existing practices of ginger cultivation

It is usually cultivated vegetation, meaning the pieces of rhizome are planted in the soil and each sprouts to form a new plant. Modern micropropagation is also being used where new plants are cloned from cells taken from a single plant [11]. The propagation through rhizome cuttings includes small bits of rhizome; each containing a bud is planted in holes or trenches 10–15 cm deep and about 30 cm apart during the months from March to May. The plants produce aerial shoots which are dried after flowering and fruiting in a span of about 8–10 months. Though ginger requires ample rainfall, there should not be any waterlogging at any time; otherwise, it may result in rotting.

In Jamaican practice, rhizomes are dug up and washed with water to remove adhered soil and dirt. Rhizomes are then peeled and washed in clean water often containing lime juice. The cleaned materials are sun-dried for 120–140 hrs; proper care should be taken that there should not be any damping or mouldy condition during the process of drying. They are then bleached by washing and again dried for 2 days.

In traditional Indian practice, the rhizome is peeled on the flat sides only followed by treatment with boiling water and sun-drying. Sometimes the dug-up rhizomes are washed in clear water and peeled or scrapped and then coated with carbonate or sulphate of lime. This protects the rhizomes from insect attack. The skin gathered from such peeled rhizomes is often distilled and oil is obtained.

Here in Sikkim, the most common practice of ginger cultivation is growing in terrace as well as in hill slopes. Ginger prefers light-textured soil which is rich in organic matter; this results in good yield of crop, and rhizomes have a better size and appearance and texture as well. Heavy soil with poor drainage should be avoided since it causes waterlogging and physiological drought.

6. Land preparation

Ginger needs fine bed preparation before sowing which is done by 2–3 ploughings or is dug 2–3 times. The sizes of the beds are variable according to available land size, but a width of 75–100 cm and height of about 15–20 cm with 30–45 cm are recommended for sowing. Raised beds should have a gentle outward slope to tackle the problem of avoid.

7. Seed selection and treatment and sowing

Healthy and disease-free planting material is the first requirement of any crop, and the same is followed in ginger too. When the crop is 6–8 months old and still green, the clumps are marked in the field. The seed rhizomes are treated with a mixture of 10% solution of bioagents, viz. *Pseudomonas* and *Bacillus*, for 30 min and dried under shade before storage or before sowing. In order to obtain good germination, the seed rhizomes are to be stored properly in pits under shade.

In Sikkim the last week of February to March is the most optimum timing of sowing; however, in most of the location, it is continued till April.

8. Sowing and spacing

Generally line sowing in shallow pits on raised beds with a row-to-row distance of 30–45 cm and plant to plant 15–20 cm is followed in ginger. Where ginger is sown as intercrop like with maize, then it is sown in 2–4 lines with spacing maintained at 60–90 cm, and maize is sown in between the rows of ginger. Usually seed rhizomes are broken into small pieces of 5–8 cm in length, weighing 30–60 quintal/ha, or even more seed is sown; however, many farmers sow full rhizome without breaking.

9. Nutrition management

Application of 20–40 tonnes of cow dung or compost in soil is recommended before sowing though heavy manuring (up to 50 tonnes/ha) in ginger field is a common practice. Farmers in Sikkim do not use any fertilizer in ginger field citing their observation that the use of fertilizers enhances rotting (soft rot), while others use dolomite which reduces rotting.

10. Mulching

To tackle the problem of post-sowing water stress, beds are covered with dry leaves, straw or other plant residues. Covering the bed with green leaves and twigs of Chillaune (*Schima wallichii*) or other forest plants is a common practice followed by some farmers. Based on the availability, mulch may vary from 5 to 20 tonnes/ha on a dry basis. Mulching not only helps in moisture management, but it also gives heat to the beds, enhances germination, reduces weeds, protects young emerging plants with heavy rain and hailstorms and minimizes soil erosion. In post-decomposition phase, mulch turns into a source of nutrients to the soil and copes up the need of growing plants.

11. Weeding and earthling up

Heavy rain is a major issue in ginger cultivation in Sikkim; it not only causes weed problem but waterlogging and soil runoff and favours outbreak of insect pest as well. To tackle the weed problem, 2–4 hand weeding is recommended as per the growth stages of the crop and weed. Weeding is followed by earthling up to cover the exposed rhizome.

12. Removal of mother rhizome (Mau)

Mother rhizome locally known as *Mau* is dug up by the month of May–June, leaving the sprouted bit of rhizome in the soil. This is done to maintain proper space to the developing rhizome and earn extra income during off season. In Sikkim, farmers get almost equal quantity, whatsoever sown, though it is of inferior quality, but being an off-season product, it fetches good market price.

13. Intercropping and crop rotation

Though ginger is grown as a sole crop in majority, it is not an uncommon thing to grow it with soybean, French bean and cowpea as intercrop. In order to provide better shade and utilize the vacant space up to maximum, farmers cultivate ginger in citrus and guava orchards as well.

Crop rotation is a common practice seen in Sikkim, and farmers tend not to cultivate ginger continuously year after year. Based on the size and suitability of the land holdings, sometimes crop rotation is followed for more than 2–4 years. The main reason of adopting crop rotation as cited by the farmers is the incidence of rhizome rot which can be minimized.

14. Issues and constraints in better production

Sikkim is a state with highly undulating topography and difficult area of reach. Farmers have small or marginal land holding which limits their choice of crops and adoption of improved technology. The steep slope and difficult terrain are other permanent issues which cause problem in field preparation, transportation of inputs and carrying of harvest to the house and markets. Winter is a dry spell in most of the areas and limits the moisture availability to crop which results in the reduced yield level, whereas during the rainy season, excess rainfall is conducive to the luxuriant growth of many weed florae which not only compete with the main crop but act as an alternate and collateral host for many insect pest and diseases. Poor infrastructural facilities and transportation issues force farmers to sale their produce at lower rates in local markets. The farmers of Sikkim are sound neither technically nor financially; this restricts the use of adequate fertilizer, improved varieties, pesticides and other inputs and recent technologies.

Sikkim is characterized by high annual rainfall (more than 2500 mm), high humidity and rather a mild and favourable range of temperature (13–23°C), most of the time cloudy, less duration of bright sunshine and great variation in altitude; a wide variety of florae and faunae give an opportunity for the survival, perpetuation and building up of many plant pathogens. The continuous rain from June–September reduces the labour efficiency drastically. It affects weeding, spraying and other agricultural operations. Sometimes severe hailstorms in April–May damage the early sown crop badly.

Due to the above reasons, this crop suffers with a number of insect pests and diseases in the field as well as in the storage causing heavy yield loss every year. Lack of knowledge, the use of poor quality seed, no crop rotation, extraction of mother rhizomes and poor storage condition are some other production constraints.

15. Insect pests

Ginger has a wide range of insects and pests affecting the crop and loss in yield. The major insect pest and their management are discussed below.

15.1 Stem borer/shoot borer (*Conogethes punctiferalis* Guen.)

This is one of the most serious pests in ginger; the infestation generally starts in June and continues till October. This pest is found more severe in Central Pendam, Sajong, Bhurung and Pachekhani areas. Shoot borer's caterpillar damages the plant by boring central shoot of the plants which in due course of time turn yellow and get dry.

Its intensity was very high (15–33%) in Zoom, Chakung, Chumbung and Malbasey areas in West Sikkim. Rouging, removing and burning the affected shoots and maintaining proper phytosanitation are the primary management strategies for this pest. Spraying of NKAE 5.0% solution or 3% neem oil at 3 ml/l of water or 0.3% of *Bacillus thuringiensis* product in an interval of 2–3 weeks during June–October is found to be effective. The process of spraying should be initiated as soon as the symptoms appear on pseudo-stem or leaves without any delay. Pruning and rouging of freshly infested pseudo-stems at fortnight intervals along with spraying are effective against the pest.

15.2 White grub (*Holotrichia* spp.)

White grub is another major pest in ginger which damages the crop by feeding upon the feeder roots and the rhizomes. The infestation of this pest is well recorded in the state and was found localized. During the August–September months of 1989, white grub was recorded an epidemic form in Bikmat village (Namthang area) of South District. The intensity of infestation was so severe that it affected 80–90% of ginger crop and forced the farmers to go for premature harvesting. The control measures include collection of adult beetles, and killing during their breeding season manages this pest as well as removal and destruction of alternate hosts like Nebhara, Gagun, Utis, Dudhilo, Bar, etc., to avoid resurrection in adult stage of pest.

16. Diseases

Though there has been the considerable number of diseases in ginger, major diseases in Sikkim conditions are classified as (i) root and rhizome diseases, (ii) foliar diseases and (iii) storage diseases and discussed below in detail. Root and rhizome diseases are very serious as they cause heavy yield loss. Among these, soft rot, yellow/dry rot, bacterial wilt and nematode diseases are the most important.

16.1 Root and rhizome diseases

16.1.1 Soft root

Soft root is one of the most devastating diseases of ginger caused by *Pythium* sp. And among them *Pythium aphanidermatum* is the most common; this disease may account the yield loss up to 50%.

Symptoms of this disease include yellowing of the tips of lower leaves which gradually spreads down to the leaf blade and leaf sheath along the margin. Initially leaves show yellowing only in the marginal area, and the middle portion remains green but later on the yellowing spreads to the whole leaf followed by drooping, withering and drying. The collar region of the pseudo-stem turns pale brown and appears as water soaked. The infected shoot can be easily pulled out from the soil; the infection from the collar spreads to the rhizome gradually. There are two stages of the disease: (a) seed rhizome stage in which the shoots arising from such rhizomes show the damping-off symptom in the seedling stage and (b) post-germination stage where the fully grown pseudo-stem withers and dies.

Soft root of ginger spreads either through infected planting material or the spores present in the soil, and rainwater assists the infestation in the second case from one field to another. Treating the seed rhizomes with 10% solution of mixture of *Pseudomonas* and *Bacillus* for 30 min followed by drying in shade before storage and the same treatment before sowing reduces the incidence of the disease.

16.1.2 Rhizome rot/dry rot/Fusarium wilt

Rhizome rot/dry rot/Fusarium wilt caused by *Fusarium oxysporum* is a common disease of ginger particularly in South Sikkim.

The disease can be identified by peculiar symptom yellowing which first appears on the lower leaves of young plants followed by yellowing involves whole of the leaf and wilting in The pseudo-stem is very weak and wilted and can be easily pulled out from the mother rhizome. The underground symptoms of the disease appear on the rhizomes and roots, showing creamy brown discoloration of vascular system. As a result of this, rotting of roots takes place and rhizome formation ceases.

16.1.3 Bacterial wilt

Bacterial wilt caused by *Pseudomonas solanacearum* was first observed in Sikkim in Rhenock and adjoining areas in an epiphytotic form causing yield loss. Water-soaked linear streaks/lesions or patches on the collar region of the pseudo-stems followed by yellowish to bronze coloration of the margin of the lower most leaf which gradually progress upwards are common symptoms of the disease. In due course of time, the leaves become flaccid showing typical wilting, yellow bronze-coloured margins and drooping. The affected pseudo-stems and rhizomes at the advanced stage of infection are slimy to touch with a varying degree of tissue disintegration, and milky bacterial exudates ooze when pressed gently. Symptoms of this disease mimics as of *Pythium*; however, when a small piece of infected rhizome is kept in water, it gives off turbid bacterial ooze, a symptom distinctly different from that of *Pythium*-infected rhizomes. *P. solanacearum*, the casual organism of bacterial wilt, grows best at 34°C.

It is difficult to manage the root and rhizome disease of ginger due to seed- and soil-borne nature; hence, intergraded disease management strategies are advised to minimize the intensity of the disease. In cultural practices, such as selection of well-drained soils for planting, avoidance of the stagnation of water, selection of disease-free planting material, removal of affected clumps and drenching of the affected and surrounding beds with 0.3% COC, drenching with 10% *Pseudomonas* and *Bacillus* helps to reduce the disease infestation. Along with the above practices, treatment of seed rhizomes with 1% Bordeaux mixture or 0.2% copper oxychloride can also be done.

16.2 Foliar disease

16.2.1 Leaf blight/leaf spot

Leaf blight/leaf spot of ginger is caused by *Leptosphaeria zingiberis* and can be identified by initial symptoms of spindle-shaped 5- to 10-mm-long yellow spots. Gradually the spots increase in size and become necrotic with ashy appearance in the centre and distinct yellow diffuse margin. The presence of light brown colour in between the central necrotic zone and outer yellow margin is another trait of the disease. Under high humid conditions, two or more spots coalesced and resulted in brightening of the affected leaves, and black dot-like minute fruiting bodies are noticed in the central necrotic portion of the affected leaves.

Spraying with 10% *Bacillus/Pseudomonas* solution or 0.2% COC at an interval of 10–15 days helps to reduce the disease intensity.

16.2.2 Phyllosticta leaf spot

Phyllosticta leaf spot is caused by *Phyllosticta zingiberi*, and peculiar symptoms are appearance of 1- to 10-mm small oval to elongated spots on leaves. The spots are usually isolated but may coalesce forming big lesions causing extensive discoloration and desiccation. A mature spot is white and papery at the centre and has dark brown margin with a yellowish hallo surrounding it.

Control measure for *Phyllosticta* leaf spot is as the same as in the case of leaf blight.

16.3 Storage diseases

Climatic conditions in Sikkim are highly favourable for fungal disease of ginger in storage conditions causing deterioration and subsequent heavy loss. In storage, *Fusarium* spp., *Botryodiplodia theobromae*, *Memnoniella echinata and Stachybotrys atra* are found to be primary fungi responsible for diseases.

To minimize or avoid disease infestation in storage, the temperature and relative humidity should be maintained at 13°C and 67%, respectively. Prestorage treatment of rhizome with 10% solution of mixture of *Pseudomonas* and *Bacillus* for 30 min dipping and there after drying in shade is helpful to reduce the incidence of the disease.

17. Recommendations for better production

- Procure ginger seed from the areas/fields where there was no disease or very less disease.
- Use always healthy (disease-free) rhizomes for sowing.
- Use always well-rotten cow dung/compost in the field.
- Follow crop rotation at least for 3–4 years.
- Avoid waterlogging in ginger field.
- Try to avoid removal of mother rhizomes as it enhances disease intensity.
- Keep the field weed-free.
- After every weeding, earthing up should be done. No rhizome should be exposed to sunshine.
- Suitable intercrop, viz. soybean, French bean and cowpea, may be practised for getting higher and higher yield.
- Take appropriate measures for disease and pest management.
- Select fully mature seed rhizomes for storage free from insect and disease attack.
- For the overall management of the insect pests and disease, in this hilly state, one should adopt as a rule integrated pest management (IPM) strategies. Sikkim is declared as an organic state and hence application of chemical to be avoided. We should encourage natural biocontrol agent and other cultural practices for the management of the pest and diseases.

- Departments/agencies should arrange regular training for the field functionaries and farers for the adoption of improved methods of ginger cultivation and plant protection measures.

- Local varieties are very promising, but their regular cultivation may pose some degeneration problem therefore. Selection of healthy and bold rhizomes should be regular practices.

18. Harvesting and yield

November to January after 8–9 months of sowing is the optimum time of harvesting the ginger; however, this follows the market demand dynamics in Sikkim. Yellowing of green leafy pseudo-stem turns followed by withering is maturity index for ginger, and crop is said to be ready for harvest. Mature rhizomes are dug carefully with a spade taking care not to bruise or break the fingers and can be stored at 13°C temperature and 67% relative humidity for the period of 10 months. Farmers normally get a yield 90–100 q/ha depending on ginger cultivation practices.

19. Processing

Ginger is extremely versatile and has a wide range of uses right from food additives to medicinal. Ginger is processed to give ginger oil, oleoresins, candy, preserves, ginger powder, starch from sport ginger, ginger brandy wine, beer and medicinal beverages and treatment for dehydration. Some of the ginger products are drained ginger, syrup ginger, dusted ginger, crystallized ginger, brined ginger, pickled ginger, dried ginger, etc. Ginger can be used in a myriad of food products ranging from bakery items to confectionary, beverages, marinades and sources, candies, ice cream and desserts, jams and spreads, prepared foods, health foils and nutraceuticals.

20. Storage

Farmers store ginger rhizomes in their houses either on floor or racks made of bamboo, while a few keep in pits and cover with paddy straw, dry leaves and soil. The bottom of the pits is covered with sand and paddy straw, and a roof is provided over it. Normally healthy and bold rhizomes are selected for storage, and frequently diseased rhizomes are sorted out to avoid spread of diseases.

21. Marketing

There is a ready market for fresh ginger in Sikkim. Before bringing the ginger to the market, farmers clean the ginger after harvest by removing adhering soil particles. Sorting is done at the trader's level to remove light, diseased, cut or deformed ginger rhizomes. In this process 8–10% produce is discarded. About 30% produce is sold in the village itself to the local merchants or commission agents. The remaining produce is taken to the market for sale to commission agents/wholesalers. The main marketing centre in Sikkim are Gangtok, Pakyong, Singtam and Rangpo in the East district; Namchi, Jorethang and Melli in the South district; Mangan and Dikchu in the North district; and Gyalshing, Rishi, Legship and Nayabazar in the West district.

22. Economic returns

Progressive farmers by adopting improved method of ginger cultivation get on an average of Rs. 150,000 per hectare (benefit-cost ratio varied from 3.50 to 3.80) provided the crop is not damaged by rhizome rot or any other devastating pest.

Author details

Vijayan A.K.[1*], B.A. Gudade[2*], Ashutosh Gautam[1], T.N. Deka[2], S.S. Bora[2], K. Dhanapal[1] and A.B. Remashree[1]

1 Indian Cardamom Research Institute, Idukki, Kerala, India

2 Indian Cardamom Research Institute, Gangtok, Sikkim, India

*Address all correspondence to: drvijayanicri@gmail.com and bgudade@gmail.com

References

[1] Ali BH, Blunden G, Tanira MO. Some phytochemical, pharmacological and toxicological properties of ginger (*Zingiber officinale* Roscoe): A review of recent research. Food and Chemical Toxicology. 2008;**46**:409-420

[2] Dedov VN, Tran VH, Duke CC, Connor M, Christie MJ, Mandadi S. Gingerols: A novel class of vanilloid receptor (VR1) agonists. British Journal of Pharmacology. 2002;**137**:793-798

[3] Hu WY, Zhang RP, Tang LP, Liu G. Research progress in the chemical constituents and pharmacology of ginger. Chinese Journal Ethnobiol Ethnomed Pharmacology. 2008;**9**:10-14

[4] Jiang H, Xie Z, Koo HJ, McLaughlin SP, Timmermann BN, Gang DR. Metabolic profiling and phylogenetic analysis of medicinal Zingiber species: Tools for authentication of ginger (*Zingiber officinale* Rosc). Phytochemistry. 2006;**67**:232-244

[5] Bhandari U, Sharma JN, Zafar R. The protective action of ethanolic ginger (*Zingiber officinale*) extract in cholesterol fed rabbits. Journal of Ethnopharmacology. 1998;**61**:167-171

[6] Fuhrman B, Rosenbat M, Hayek T, Coleman R, Aviram M. Ginger extract consumption reduces plasma cholesterol, inhibits LDL oxidation and attenuates development of atherosclerosis in atherosclerotic, apolipoprotein E-deficient mice. The Journal of Nutrition. 2000;**130**:124-131

[7] Nicoll R, Henein MY. Ginger (*Zingiber officinale* roscoe): A hot remedy for cardiovascular disease. International Journal of Cardiology. 2009;**131**:408-409

[8] Shukla Y, Singh M. Cancer preventive properties of ginger: A brief review. Food and Chemical Toxicology. 2007;**45**:683-690

[9] Babu MS, Kumar BP, Swami DV, Krishna KU, Emmanuel N. Performance of ginger (*Zingiber officinale* Rosc) varieties under shade net condition of costal Andhra Pradesh. International Journal of Current Microbiology and Applied Sciences. 2017;**6**(7):494-498

[10] Chhetri P, Gudade BA. Organic production of ginger: A prominent sub-Himalayan spice. Popular Kheti. 2013;**1**(4):207-210

[11] Bhatt N, Waly MI, Musthafa ME, Ali A. Ginger: A Functional Herb. New York, USA: Nova Science Publishers, Inc.; 2013. ISBN-978-1

Chapter 5

Diseases of Ginger

Gupta Meenu and Tennyson Jebasingh

Abstract

Ginger is one of the earliest known oriental spices grown for its edible rhizome, which is widely used as a fresh vegetable, spice, and as a popular folk medicine. Ginger crop is being affected by insect pests, and pathogenic and non-pathogenic diseases cause production constraints. Severely, various pathogenic diseases of viral, bacterial, fungal, and nematode origin reduce its potential yields drastically. Among the various diseases, soft rot, yellows, *Phyllosticta* leaf spot, storage rot, bacterial wilt, mosaic, and chlorotic fleck are important. The present chapter includes the symptoms, causative agent, disease cycle, epidemiology and host resistance, cultural, biological, chemical, and integrated management of these diseases.

Keywords: ginger, soft rot, yellows, leaf spot, wilt, mosaic, chlorotic fleck

1. Introduction

India is considered as a '*magical land of spices*' with diverse variety of spices. Ginger (*Zingiber officinale* Roscoe) is earliest known oriental spices, belonging to the family, *Zingiberaceae*. Though entire plant is refreshingly aromatic, the underground rhizomes of this crop are valued as spice. It is one of the commonly consumed dietary condiments in the world and has high medicinal properties.

Ginger is being cultivated in the various parts of the world. The total production of ginger in the world is 1683 thousand tons with the total acreage of 310.43 thousand ha [1]. China, India, Nepal and Thailand are the major producers of ginger in the world. India is the leading producer and exporter of ginger in the world. Annually, India produces 385.33 thousand tons of ginger [1].

The production of ginger is being affected by biotic and abiotic factors. Biotic factors include virus, bacteria, fungus and nematodes [2, 3]. Among the biotic factors, bacteria are most important, causes wilt and soft rot. Fungus is the next major pathogen causes rhizome rot, soft rot, Sclerotium rot and yellows disease. Nematode produces root knot disease and virus's causes mosaic and chlorotic fleck in ginger plants reduce the yield of the rhizome. The ginger is also affected by various insects like *Conogethes punctiferalis*, *Aspidiella hartii*, rhizome scale, rhizome fly and thrips. Abiotic factor causes sunburn (due to high light intensity) and lime-induced chlorosis (due to excessive liming in soil) in the ginger crop.

It is imperative to know the symptoms of the disease, the organism responsible and the protection measures. This chapter emphasizes the importance of diseases of ginger by including the symptoms, causative agent, epidemiology and protection from these diseases.

2. Viral disease

2.1 Mosaic

2.1.1 Symptoms

The symptoms appear with yellowish and dark-green mosaic on leaves of ginger in the early stage and stunted of leaves and rhizomes at the late stage of infection. Infection of this virus on ginger causes severe reduction of rhizome yield.

2.1.2 Causative agent

Virus isolated from affected ginger plants by So [4] named the virus as Ginger mosaic virus. The virus infected ginger possesses spherical particles with a diameter of 23–38 nm. The purified virus particle showed positive to serological reaction for cucumber mosaic virus (CMV) antiserum. Serological relationship and electron microscopic observation of this virus conclusively suggested that ginger mosaic virus could be the CMV group. To support this, last year (2018), a group from Malaysia identified that the mosaic disease in ginger was due to the CMV which is based on the partial nucleic acid sequence of coat protein (GenBank: MH355647.1).

2.1.3 Epidemiology

Virus produces mosaic disease in 18 cultivars of ginger and 23 other plant species [4]. The virus is transmitted through the sap to different plants which are believed to the hosts of CMV [4]. Nambiar and Sarma [5] failed to absorb the sap transmission from ginger to ginger, ginger to *Nicotiana tabacum* var. Harrison special, *N. tabacum* var. xanthi, *N. tabacum* var. rustica, *N. glutinosa*, *Elettaria cardamomum*, *Curcuma longa*, and *C. aromatica*. Assessing the concentration of virus in various parts of ginger revealed that leaves and flower possess higher concentration of virus than rhizome, stem and other parts of the plant.

2.1.4 Transmission

The virus is transmitted by insect vectors such as *Myzus persicae*, *M. certus*, *M. humuli*, *Macrosiphum euphorbiae* and *Rhopalosiphum insertum*. *Myzus persicae* and *M. certus* are comparatively more efficient vectors for this virus.

2.1.5 Protection

Hot-water and hot-air treatments of affected rhizomes at 45 and 50°C for 3, 6, and 12 h does not alleviate symptoms [5]. Ginger mosaic virus in standard extract is inactivated with 10 min exposure at 60°C [4].

2.2 Chlorotic fleck

Thomas [6] from Australia reported the chlorotic fleck in ginger is because of ginger chlorotic fleck virus (GCFV). From Australia, this virus is distributed to a number of countries which import ginger from Australia.

2.2.1 Symptoms

The leaves of infected plants show chlorotic flecks, 1–10 mm long on the centered and parallel to the veins (**Figure 1**). Symptoms started appearing in the young leaves at the 3–4 weeks of infection and subsequently to other leaves. No obvious symptoms occur on the rhizomes.

2.2.2 Causative agent

Chlorotic fleck in ginger is due to ginger chlorotic fleck virus (GCFV). It can be readily purified from the virus ginger leaves through ultracentrifugation with a sedimentation coefficient of 111 s. The purified virus is an isometric nature with size of 28–33 nm. Virus particle contains a major coat protein with molecular weight of 29 kDa and single stranded RNA with molecular weight of 1.5 × 106 Da.

2.2.3 Epidemiology

It differs from mosaic virus in particle properties, host range and serology. Many properties namely size of particles, possession of ssRNA, salt-labile nature of particles and a limited host range of this virus are similar to sobemovirus group [7] but it, serologically unrelated to several sobemoviruses including lucerne transient streak virus, cocksfoot mottle virus, sowbane mosaic virus, Solanum nodiflorum mottle virus, southern bean mosaic virus, velvet tobacco mottle virus and turnip rosette virus.

Figure 1.
Chlorotic fleck symptoms on ginger leaf (source: https://www.plantwise.org).

2.2.4 Transmission

GCFV is mechanically transmitted only to ginger but not like ginger mosaic virus, transmitted by *Myzus persicae, Pentalonia nigronervosa* and *Rhopalosiphum maidis* or *R. padi*.

2.2.5 Protection

The viral diseases of ginger are controlled in tissue cultures by heating at 50°C for 5 min [8].

2.3 Viruses infecting flowering ginger

Flowering ginger is a member of the Alpinia genus of *Zingiberales* order of *Zingiberaceae* family. Banana bract mosaic virus (BBrMV) (Potyvirus genus of the family *Potyviridae*), Canna yellow mottle virus (CaYMV) and Banana streak virus (BSV) of Badnavirus genus of the *Caulimoviridae* family are reported to infect flowering ginger [9].

3. Bacterial diseases

3.1 Bacteria wilt

Bacterial wilt of ginger is the most serious rhizome-borne diseases. It is also soil and seed-borne disease. It is widespread and exceedingly destructive for the ginger grows in tropical, subtropical and warm temperate regions of the world. Bacterium causes rapid wilt in ginger within 5–10 days of the infection [10]. The severity of the disease is occurred due to the rapid spread of the pathogen happens during the favorable environment conditions like high rain fall and warm weather.

3.1.1 Symptoms

Initially, water soaked patches or linear streaks appear at the collar region of the pseudo stem and then progresses both upwards and downwards. The pseudo stems from the infected plant can be easily separated with a gentle pull and can be broken off at the base. Mild drooping and curling of leaf margins of lower leaf is the first prominent symptom occurred after the infection, then the infection spread upwards later. Yellowing starts from the lower-most leaves which gradually progresses upwards. In the advanced stage, infected ginger exhibit intense yellowish and wilting symptoms (**Figure 2**). Dark streaks are observed in the vascular tissues of the affected pseudo stems. In the infected plants, leaf sheaths look yellowish to dull green. The leaves roll up and the whole plant dries up, finally. The plants which are infested by the disease stand persistently and do not collapse. The base of the infected pseudo stem and the rhizome emit foul smell. The affected pseudo stem and rhizome extrudes milky ooze from the vascular strands while they pressed gently. Milky bacterial exudates ooze out from the cut end (**Figure 2**).

3.1.2 Causative agent

Ralstonia solanacearum Yabuuchi causes bacterial wild disease in ginger. It is considered as the second most important destructive bacterial pathogens identified

Figure 2.
Yellowing of leaves and oozing of rhizome due to bacteria wilt disease (source: Vikaspedia).

to date. Four biotypes of *R. solanacearum* have been identified so far. Biotype III from India causes the wilt in ginger [11]. Biotype III of the bacterium cause slow wilt and biotype IV causes rapid wilting and death [12]. Biotype III is restricted to ginger plant and its weeds whereas biotype IV infects a wide host range including potato, tomato, eggplant, *Capsicum frutescens*, *Zinnia elegans* and *Physalis peruviana*. Biotype II is reported to infect only potato plants [13]. In Indonesia, the race 1 of biovar III is responsible for bacterial wilt in ginger [14]. Nematodes in the soil increase the incidence of wilt in ginger [15]. Nelson [16] observed that the host range of race 4 of biovar III of *R. solanacearum* is restricted to edible ginger. Twelve of 14 species of ginger belonging to Zingiberaceae and Costaceae are highly susceptible to all the strains of race 4 and susceptible plants wilted within 21 days [17].

Yu et al. [18] studied the genetic diversity of *R. solanacearum* strains collected from ginger plants growing in Hawaii island with amplified fragment length polymorphisms (AFLPs) showed a high degree of similarity between the strains with the value of 0.853. Strains from ginger in Hawaii showed less similarity for the strains from tomato (race 1) and heliconia (race 2) [18]. Kumar and Sarma [10] have characterized the isolates of *R. solanacearum* on the basis of their membrane protein pattern on SDS PAGE and serological detection (NCM-ELISA) using *R. solanacearum* specific antibodies. Detection of *R. solanacearum* from rhizomes and soil through PCR was developed for bacterial wilt [19–21]. Shan et al. [22] reported the complete genome sequence of *R. solanacearum* SD54, a race 4 biovar 4 (R4B4) strain from a diseased ginger plant.

3.1.3 Transmission

R. solanacearum is a soil-borne disease spreads from the soil by adhering to hands, boots, tools, vehicle tires and field equipment; through water during irrigation and rainfall; and through infected ginger rhizomes [23]. From the soil, this bacterium enters into the ginger plant through the roots and rhizomes *via* the openings where lateral roots emerge or through wounds caused while handling and by root-knot nematodes or parasitic insects [24]. The bacteria survive in the infected plant debris and as free-living in soil. Ginger crops on this infested soil severely affected and completely lost to the disease [16].

3.1.4 Protection

Difficulties are associated with controlling this pathogen due to its endophytic nature, survive in deep soil, travel along water, and its relationship with weeds. The available physical, chemical, biological, methods and cultural practices to manage this pathogen are discussed below.

3.1.4.1 Physical methods

Physical control methods like solarization and hot water treatment have proved to be effective against the pathogen. Rhizome solarization for 2–4 h on ginger seeds reduce the bacterial wilt (90–100%) at 120th day of planting, and further with discontinuous microwaving (10-s pulses) at 45°C reduces the wilt by 100% [25]. Tsang and Shintaku [26] reported that bacterial wilt pathogen is eliminated when the rhizome is exposed to heat for 30 min at 50°C and 45 min at 49°C, respectively. Exposing ginger seed pieces to hot air at 75% RH until their temperatures attained 49°C for 30 and 60 min and 50°C for 30 min, resulted in minimal injury to the hosts. More than 87% of the seed pieces germinated without adverse effect on growth.

3.1.4.2 Chemical methods

Treating seed rhizomes with emisan in addition to plantomycin for 30 min followed by three sprayings, first at 30 days after planting and others at an interval of 15 days, gave good protection against wilt disease [27]. Streptocyclin (20 g/100 l water) treatment on seed and drenching the soil with 0.2% copper oxychloride, protect the seed from the bacteria infection. Sinha et al. [28] observed that streptomycin and streptopenicillin are superior over the other antibiotics against the bacterial wilt pathogen.

3.1.4.3 Biological methods

Bacillus subtilis strain 1JN2, *Myroides odoratimimus* 3YW8, *B. amyloliquefaciens* 5YN8, and *Stenotrophomonas maltophilia* 2JW6 are used as biocontrol showed efficacies greater than 50% against bacterial wilt of ginger [29].

3.1.4.4 Cultural practices

The effective management of ginger against the pathogen is depends up on the various factors. Selecting the disease free seeds, sowing the disease free seed on disease free land based on previous history, following 4–5 years of crop rotation with non-solanaceous plants, planting on raised beds (help to avoid water stagnation during rainy season), giving thick mulching (to avoid weed growth and to conserve soil moisture), reduces the disease causing potential of the soil. Indrasenan et al. [30] suggested selection of healthy seed rhizomes, eradication of weeds and adoption of an effective crop rotation as control measures for the disease.

3.2 Bacterial soft rot

This is not considered as a major problem in ginger, but periodic outbreaks occur when ginger is planted in waterlogged soil [12]. The disease is more prevalent in rhizomes that have formed deep in the ground. The sections closer to the surface are generally healthy. Disease is not found in well-drained soils.

3.2.1 Symptoms

The disease causes a gradual softening of the rhizome tissue accompanied by an offensive odor.

3.2.2 Causative agent

Erwinia chrysanthemi is the only *Erwinia* species responsible soft-rot of ginger. High temperatures, saturated soils and injury during seed preparation all tend to exacerbate the disease.

3.2.3 Disease cycle

Ginger can have the bacteria either from the infected seed, or from direct inoculation, through wounds or natural openings. The bacteria started to feed liquids released from injured cells and multiply. Bacteria secrete pectolytic enzymes degrade and break the cells providing more food for the bacteria. Often the epidermis is left unscathed, keeping the rotten flesh contained within until a crack allows the ooze to leak out and infect others around it. The bacteria from the harvested infected plant to others placed with it and also through the insects.

3.2.4 Prevention

The most effective way to prevent this disease is simply keeping sanitary growing practices. It includes removing all plant debris from storage ware houses and disinfecting walls and floors with either formaldehyde or copper sulfate between harvests, maintaining low humidity and temperature of the storage facility with an adequate ventilation system. It also by planting in well-drained soils, rotating susceptible plants with non-susceptible plants.

4. Fungal diseases

Ginger plantation is majorly affected by deuteromycetous group of fungi cause variable symptoms [31]. In India, fungal diseases reduce the potential yield to a greater extent in field, storage and market and may cause losses of even more than 50% [32].

4.1 Soft rot/rhizome rot

Soft rot found in all the ginger growing countries, reported as the most dangerous and destructive disease of ginger which can reduce the production by 50–90%. Disease cause significant losses during warm and humid conditions. Butler on 1907, recorded this disease first time in Surat district of Gujarat, India [33].

4.1.1 Symptoms

This disease is prevalent in ginger crop throughout the growing period. Sprouts, roots, developing rhizome and collar region of the pseudo stem are highly prone to infection. Symptoms first appear on the aerial parts of the plant. Pathogen form watery and brown lesions in the collar region of the pseudo stem. Later the lesion enlarges, coalesce and cause the stem to rot and collapse [34]. In the old leaves, initially, yellowing (chlorosis) symptoms appear in the tips, which then spread downward along the margin involving the rest of the leaf blade and, eventually, the leaf sheath. Later, chlorosis from the older leaves progress to younger leaves start developing a similar symptom progression until the entire plant dies [35] (**Figure 3**). The appearance of lesion in pseudo stem and chlorosis in the leaf indirectly show the sign of rhizome rot. Due to the infection, rhizomes appear soft, brown, water soaked, rotten, and decay gradually [34] (**Figure 3**). It is not like bacterial rots, the soft rot caused by fungus does not produce offensive odors.

4.1.2 Causative agent

Eleven species of *Pythium* are responsible for the soft rot diseases in ginger are identified so far. Among the 11 species, *P. myriotylum* and *P. aphanidermatum* cause severe damage in warm climates. *Fusarium* is another fungus reported to cause soft rot of ginger. Among different species of *Fusarium*, *F. oxysporum* causes decaying of ginger rhizomes [36].

Booster PCR method to detect *P. myriotylum* from the infected ginger rhizomes has been established [37]. Yella *et al.* [38] has developed a simple technique to produce oospores in *P. myriotylum*. Le et al. [39] developed PCR-RFLP based detection of isolates of *Pythium* by studying the genetic variation between *P. myriotylum*.

4.1.3 Disease cycle and epidemiology

It is seed and soil borne disease carried by two ways: (1) Through diseased rhizomes scales [40] and (2) through soil as oospores. *Pythium* survive in plant debris as perennating oospores, an important source for infections. The favorable conditions like wet soil conditions, high soil moisture and soil temperature influence the development of the oospore [41]. Severity of disease is influenced by high rainfall and when rhizomes planted in heavy clay soil with poor drainage. Temperature, 34°C is optimum for the germination of *P. aphanidermatum* and *P. myriotylum* spores.

4.1.4 Protection

Soft rot is a complex disease problem and various methods should be applied to combat the disease.

Figure 3.
Yellowing of leaves and rotting of rhizome due to soft rot disease (source: Vikaspedia).

4.1.4.1 Physical methods

Using disease-free seeds is an essential step to prevent the contamination of *Pythium* [41]. The various strategies like seed protection (biologically, physically or in combination), seed disinfestations (to control spores on the seed surface), seed disinfection (to eliminate pathogens living inside the cells) such as chemical infusion protect the seed from the infection as well as improve the germination of seeds [42]. In order to protect the plant from the persistence nature of the pathogen in soil, crop rotation with rice and corn (tolerant to pathogens of ginger) can be done to minimize the recurrence of pathogen in the next harvest [42–44]. Using soil with higher clay content and lower pH, suppress *P. zingiberum* and *F. oxysporum* [45]. Soil solarization with biological control methods is beneficial to plant growth and antagonistic to many pathogens like fungi, bacteria, nematodes, and arthropods. Phytosanitation is needed when the symptoms of soft rot appear in the ginger plantations.

4.1.4.2 Chemical methods

Pythium spp. is soil borne pathogen survive in the soil for a long period [46]. It is very difficult to control this. Treating the rhizomes with mancozeb, ziram, guazatine, propineb and copper oxychloride for 30 min effectively control this pathogen [47, 48]. Treating seed with Ridomil MZ (1.25 g/L) increase the survival of rhizomes (about 30%) in comparison to hot water treatment (at 51°C for 30 min) [49]. Rajan et al. [50] reported that rhizomes are protected from *Pythium* while treating the seed with Fytolan (copper oxychloride) (0.2%), Ridomil (500 ppm), Bavistin (carbendazim) (0.2%) and Thimet. Application of fungicides such as metalaxyl, Ridomil, Maxam XL (fludioxonil) and Proplant (propyl carbamate hydrochloride) on seed give significantly better result for *P. myriotylum* than carbendzim alone [51]. Soil drenching with Zineb, captafol, methyl bromide, mercuric chloride, thiram, phenyl mercury acetate, copper oxide and mancozeb protect the ginger crop from soft rot disease [52]. Application of Metalaxyl in soil is effective to control the soft rot [53]. Treating seed with a mixture of metalaxyl and captafol and treating the soil with the same chemicals after 3 months of planting controlled the disease [54]. Metalaxyl formulations (Ridomil 5G and Apron 35 WS) in soil and treating the seed gave very good protection against soft rot [55]. Srivastava [56] effectively managed *P. aphanidermatum* by drenching Zineb or Mancozeb in the soil after treating the rhizome with carbendazim. Treating with 1% formaldehyde and then planting the ginger under shade had lower incidence (19.4%) of soft rot as without shade (41.3%) [57].

4.1.4.3 Biological methods

Trichoderma spp. is the best biocontrol agents for soft rot. Non-volatile and volatile compounds from *T. viride* inhibit the growth of *P. myriotylum* [58]. Similarly, *T. harzianum* and *T. saturnisporum* also showed strong antagonism against *P. splendens* [59]. Several rhizobacteria were reported to possess strong antagonistic activity on *P. myriotylum* growth [60]. Growth of *P. aphanidermatum* on potato dextrose agar (PDA) was completely inhibited when it amended with onion and garlic extracts at 5 and 7.5% (v/v), respectively [61]. Rakesh et al. [62] showed inhibition of *P. aphanidermatum* with treatment of fresh and stored cow urine at a concentration of 20% (v/v). Application of *Boerhavia diffusa* leaves and seed powder of *Azadirachta indica* in the soil before the planting reduces the infection [63]. Adding neem cake in the soil also found to suppress *P. aphanidermatum* [64]. Application of *Jeevatu* based organic liquid manure in the field, control the soft rot and the spreading of this disease [65]. Coating the seeds with *Trichoderma* spp. reduce the soft rot, 2–3 times less than that of the untreated control [66]. Reduction of soft rot occurred when seed was first disinfested with HOCl (1%) followed by soaking in *Trichoderma* spp. and followed by three applications of talc-based formulation (3×10^6 CFU/g) to the soil at 15 day intervals from the time of planting [67]. Treating with *T. harzianum*, *Glomus mosseae* and fluorescent Pseudomonad strain G4 together inhibit the infection up to 10% than treating individually with 30, 43, and 50% infection, respectively [68].

Treating with *Burkholderia cepacia* and *T. harzianum* showed production efficiency of 84% with reduction of disease incidence of 79.7% [67]. Extracts from *Jacaranda mimosifolia* and *Moringa oleifera* gave the best inhibitory effect of 27.7% against *P. aphanidermatum* [69].

4.1.4.4 Cultural practices

Cultural practices such as seed selection, crop rotation, organic amendment, tillage, drainage and quarantine practicing in ginger plantation not only control the disease but also limit the spread of *Pythium* spp. [70].

Planting the disease free rhizomes is the best method to manage the disease [71]. Harvey and Lawrence [72] believed that crop rotations reduce *Pythium* spp. populations. Rames et al. [73] found that fungal and bacterial populations were significantly greater in soil when pasture grass (*Digitaria eriantha* subsp. *pentzii*) grown continuously than the soil is treated with fumigant or left as bare fallow.

Addition of organic matter from various plants, oil seed cakes and neem cake, reduce the incidence of soft rot [74, 75]. Mixing organic matter with poultry manure and sawdust enriched the soil microbial populations in the ginger growing soils [73] and enhance the soil carbon levels and water infiltration rates, support the growth and yield of ginger [51, 76]. Kadam et al. [77] reported that neem seed cake with least average mortality (20.3%) followed by poultry manure (22.7). Kumar et al. [78] also reported that *Schima wallichii* and *Datura* spp. were the best mulches with respect to inhibit soft rot caused by *P. aphanidermatum*. Smith and Abbas [51] recommended proper water drainage in the ginger filed to inhibit the *Pythium* spp. zoospores germination and spreading.

4.1.4.5 Host resistance

Identifying *Pythium* resistant ginger is an ideal for effective soft rot disease management. Indrasenan and Paily [79] identified Maran cultivar resistant against *P. aphanidermatum*. Setty et al. [80] identified Supraba and Himachal Pradesh cultivars showing less than 3% soft rot incidence. Senapati and Sugata [81] found one resistant cultivar and eight others with moderate resistance while screening 134 ginger varieties available in Koraput, Orissa, India. Kavita and Thomas [82] reported that *Zingiber zerumbet* accession is suitable candidates to obtain soft rot resistance for ginger. Screening 650 ginger accessions by Bhai et al. [83] found that only 7% of the accessions have the relative resistance to the *Pythium* sp.

4.1.4.6 Integrated management

Following a single approach is not ideal way to a have soft rot resistant. It is an important to have a multiple approach. Smith and Abbas [51] focused on cultural practices with strict quarantine procedure to manage the disease. Soil solarization with fungicides effectively reduces the *P. myriotylum* [84]. Treating rhizomes with Ridomil MZ (metalaxyl and mancozeb) and soil drenching with Thimet (Phorate) and Ridomil MZ after planting gave the best control of *P. myriotylum*. Seed solarization at 47°C for 30 min within a polyethylene sheet is ideal for *Pythium* sp. disinfestations [85]. Having soil solarization for longer periods reduces *Pythium* spp. populations with lower disease incidence [86]. Applications of bioagents such as *T. harzianum*, *P. fluorescens* and *B. subtilis* together reduce the disease than individually on the rhizomes (8.64%) as well as on tillers (12.50%) [87]. Rhizome treatment with copper oxychloride followed by neem extract, suppresses the soft rot disease [88]. Rhizome treatment in hot water at 47°C for 30 min and drenching of soil *T. harzianum*, followed by mancozeb are most effective in inhibiting *Pythium* sp. as well as improve the yield of ginger [89].

4.2 Yellows/wet rot

Yellows disease is serious problems of ginger causes stem and rhizome rot. It is wide spread and prevailed in warm and humid environmental conditions. It was first described by Simmonds [90] from Queensland. Later this disease was reported from Hawaii [91] and India [92].

4.2.1 Symptoms

Yellowing starts on the margins of the lower leaves which gradually spreads and cover the entire leaves. Later, the yellowing diffuse to older leaves. Old leaves dry first and then younger leaves. The affected plants wilt and dry up but do not fall on the ground in contrast to soft rot and bacterial wilt. Infected pseudo stem comes off from the rhizome with a gentle pull. Rhizomes become soft and watery with a creamy discoloration of the vascular system and cortical rot. Plants may show a premature drooping, wilting, yellowing, drying in patches or in whole bed and show stunting. Rotting of roots is common. Mycelial growth in the form of white, peach or buff colored cushions can be seen on the surface of rhizomes [93].

4.2.2 Causative agent

Fusarium oxysporum f.sp. *zingiberi* Trujillo is a causative agent for yellow disease [94]. Other species of *Fusarium* such as *F. solani* (Mart.) Sacc., *F. equiseti* (Corda) Sacc., *F. graminearum* Schwabe, *F. moniliforme* Sheld and some unidentified *Fusarium* spp. were also reported to be associated with yellow disease ([95–97]. *F. solani* is the second most important species [96, 98]. Isolates of *F. oxysporum* f. sp. *zingiberi* differed in their aggressiveness [99].

Genetic variation of *F. oxysporum* f.sp. *zingiberi* isolates with DNA amplification fingerprinting (DAF) segregated the isolates into three haplotypes based on 17 polymorphic bands generated with five primers [100]. Two haplotypes showed very little genetic variation (98.6% similarity), whereas the third haplotype was quite distinct in terms of its molecular profile (77.2% similarity). Shanmugan et al. (2013b) studied genetic variability of 32 *Fusarium* isolates from diseased ginger rhizomes from Western Himalayas in India. They were analyzed by the unweighted pair group method with arithmetic averaging using randomly amplified polymorphic DNA amplicons. Of two major clusters formed, one was dominated by *F. oxysporum* and the other by *F. solani*. Morphological, cultural, pathological and molecular variability among *F. oxysporum* f.sp. *zingiberi* isolates were studied by Gupta et al. [101]. Molecular variability revealed 0 to 80% variation among 19 isolates and they were grouped into two different major groups each comprising of 10 and 9 isolates, respectively.

4.2.3 Disease cycle and epidemiology

The seasonal carryover of fungus inoculum takes place through infected rhizomes and soil. The fungus survives in soil as chlamydospores which may remain viable for many years in the field. The fungus spreads through infected seed rhizomes and about 87% of field infection is due to infected rhizomes [102]. The secondary spread of the disease can also take place through irrigation water and by mechanical means.

For the development of yellows disease, a temperature range of 15–30°C is favorable (the optimum being 23–29°C) with very high humidity and continuous presence of free water [103]. Maximum disease incidence occurred when soil temperature ranged from 24 to 25°C and the soil moisture from 25 to 30% [104].

4.2.4 Protection

4.2.4.1 Chemical methods

The various chemicals have been shown promising result against the pathogen [105–109]. So far fungicides like Bavistin 50WP, Ridomil Gold MZ-72, Captan, Dithane M-45, copper oxychloride and Bordeaux mixture are found to be effective against the disease [110, 111].

4.2.4.2 Biological methods

Microorganisms like *Trichoderma harzianum* and *Gliocladium virens* [112], *T. viride* and *T. harzianum* [113], *T. viride* [114] inhibits *F. oxysporum*. Among six *Streptomyces* species, SSC-MB-02 was effective against *F. oxysporum* [115]. Treatment of *Bacillus cepacia* and *T. harzianum*, increased the production of rhizome as well as protect the plant from yellow disease [59]. Talc-based formulations with rhizobacteria strain XXBC-TN (*Bacillus subtilis*) and a mixture of S2BC-1 (*B. subtilis*) and TEPF-Sungal (*Burkholderia cepacia*), inhibit *F. oxysporum* and *F. solani*. Treatment resulted in increased rhizome production with reduced yellows. Increased protection is due to increase in defense enzymes such as chitinase, β-1,3-glucanase and polyphenol oxidase and promote plant growth and rhizome [67].

Among 14 plant extracts, the plant extract of *Ferula asafoetida* (68.51%) has high protection against *Fusarium*, followed by *Ocimum* leaf extract (60.16%) [116]. Leaf extracts of *Swietenia macrophylla* King, *Azadirachta indica* A. Juss., *Hyptis suaveolens* (L.) Poit., *Polyalthia longifolia* (Sonn.) Thw., *Boerhavia repens* L. var. *diffusa* (L.) Hook. and *Tithonia diversifolia* A. Gray showed 100% resistance against *F. solani* [117]. Among the various nanoparticles synthesized from plants, a sulfur nanoparticle has high inhibitory effect on *F. oxysporum* [118]. Further, sulfur nanoparticles with fungicides like bavistin, ridomil gold, sunflex and streptocycline were evaluated for the inhibition of the fungus showed that sulfur nanoparticles with bavistin with high inhibition.

4.2.4.3 Cultural practices

The disease is spread mainly through contaminated rhizomes. Planting healthy seed rhizomes is the best way to avoid this disease [71, 119]. Applying organic manure, tillage and crop rotation reduces the yellows disease [76]. Planting the seed rhizome (size of 50–75 g) with spacing of 25 × 30 cm is ideal for good yield and lower disease incidence [120].

4.2.4.4 Host resistance

Developing a *Fusarium* resistant variety would be ideal for effective yellows disease management. Two cultivars: SG 666 [121] and Kerala local [122] are reported to have disease resistant. The resistance is due to the presence of a resistance (R) gene of CC–NBS–LRR [123].

4.2.4.5 Integrated management

The integrated approach like treating the seed and well as in soil with fungicides mancozeb and carbendazim and biocontrol agents like *T. harzianum, T. hamatum and G. virens* reduces the incidence of disease [124]. Hasnat et al. [110] reported lowest yellows disease while treating with Ridomil Gold along with poultry waste, Bavistin 50WP, Dithane M-45 and saw dust in the soil at 240th days of planting.

4.3 Leaf spot

Leaf spot disease is becoming increasingly important in many places of India due to severe leaf rot and blight it causes. Ramakrishnan [125] reported this disease first time from Godavari district of Andhra Pradesh and Malabar area of Kerala.

4.3.1 Symptoms

On the young leaves, small spindle to oval to elongated spots size of 1–10 mm × 0.5–4 mm appears. Later, the spots developed as white papery center and dark brown margins surrounded by yellow halos [125]. The spots increase in size and coalesce to form larger lesions which lead to the reduction of effective photosynthetic area on the leaves. The affected leaves become shredded and may suffer extensive desiccation (**Figure 4**). Symptoms appear first on younger leaves. As the plants put forth fresh leaves, these get infected subsequently.

4.3.2 Causal organism

Phyllosticta leaf spot is caused by *Phyllosticta zingiberi* T.S. Ramakr. The fungus forms amphigynous, subglobose, dark brown ostiolate pycnidia on the host measuring 78–150 μm in diameter. On standard media, the fungus forms pycnidia having 100–270 μm diameter bearing hyaline, unicellular, oblong, big guttulate spores measuring 3.7–7.4 × 1.2–2.5 μm [125].

4.3.3 Disease cycle and epidemiology

Primarily, disease is spread from the debris of infected plants and from the infected seeds. Under the laboratory condition, pycnidiospores and mycelia of fungi alive for 14 months [126] and spores remain viable in soil even at 25 cm depth for 6 months. Pycnidia of *Phyllosticta* survive in the leaf debris even during the summer. Mycelium of *Phyllosticta* grows at 25.0–27.5°C with maximum and minimum of 32.5 and 10.0°C, respectively. The mycelial growth was inhibited at 5 and 35°C [127].

During the rainfall, the dispersal of spore occurred. Higher precipitation along with high wind give greater impact on dispersion of spores to many leaves which are

Figure 4.
Symptoms of spot on ginger leaf (source: Vikaspedia).

in longer distances [128]. Factors like air temperature, relative humidity and rainfall influence the incidence of disease to an extent of 85.5% [129]. Six to seven months old plants are prone to infection of *Phyllosticta* and 2 weeks old leaves are most susceptible. Temperature of 23–28°C with intermittent rain favored the occurrence of this disease. Senapati et al. [130] observed that incidence of leaf spot disease was less for the plants grown under partial shade or as intercrop in coconut. Incidence of disease is higher and yield of rhizome is reduced when the ginger is cultivated continuously without the rotation of the crop [131].

4.3.4 Protection

4.3.4.1 Chemical methods

Treating the plant with Bordeaux mixture, zineb and maneb are effective in treating the disease [132]. Grech and Frean [133] observed that spraying mixture of benomyl (0.1%), mancozeb (0.2%) and soluble boron (0.1%) and iprodione (0.2%) reduces the production of disease. Highest reduction of the incidence of disease is observed with spraying chlorothalonil [127]. Verma and Vyas [134] observed higher protection while spraying carbendazim (0.15%) and mancozeb (0.25%) and due to this higher yield also obtained. Increased yield of rhizome and decrease disease incidence was found while treating the rhizome and doing foliar spraying with Bordeaux mixture (1%), Companion (0.2%), Indofil M-45 (0.25%), Unilax (0.2%) and Baycor (0.05%) [129].

4.3.4.2 Cultural practices

Growing the crop under the partial shade reduce the severity of Phyllosticta leaf spot. Growing the ginger under the partial shade of mandarin orange increase the growth of the plant and reduce the disease intensity [135]. Reduction of leaf spot and sun burn on leaves occurred while growing plants under shade have increased the number of tillers per clump recommends the growing of ginger in partial shade to avoid the fungicidal spray [107].

4.3.4.3 Host resistance

None of the 18 cultivars tested in Karnataka, India were resistant to *Phyllosticta* leaf spot [136]. However, the cultivars Narasapatom, Tura, Nadia, Tetraploid and

Thingpani were classed as moderately resistant with a disease index less than 5%. In Himachal Pradesh, India none of the tested material of ginger was rated resistant to *P. zingiberi,* however, eight lines showed moderate resistance [137]. Different workers obtained variable results and none of the tested cultivars showed high degree of resistance [137, 138]. Nageshwar et al. [138] screened 100 accessions of ginger for their reaction and tolerance to leaf spot under field conditions and of them, 11 accessions were found tolerant and further 42 were moderately tolerant. Senapati et al. [130] found that PGS-16, PGS-17 and Anamica as moderately resistant out of 135 ginger cultivars tested.

4.4 Storage rots

Post-harvest losses in ginger are a serious concern. The post-harvest losses are affected by various biotic and abiotic causes. Rhizomes are stored for seed and commercial purpose. During storage, rhizomes soft are affected by fungi [71] and bacteria.

4.4.1 Symptoms

Fungal mycelia discolored the surface of rhizome accompanied with dry rotting and decaying (**Figure 5**).

4.4.2 Causal organism

Fungus like *F. oxysporum* Schlechtend ex Fr., *P. deliense* Meurs and *P. myriotylum* Drechs. [139], *Geotrichum candidum* Link [140], *Aspergillus flavus* Link ex. Fr. [141], *Cladosporium tenuissimum, Gliocladium roseum* Bainer, *Graphium album* (Corda) Sacc., *Mucor racemosus* Fresen., *Stachybotrys sansevieriae, Thanatephorus cucumeris* (Frank) donk and *Verticillium chlamydosporium* Goddard [142]. *Pythium ultimum, Fusarium oxysporum* and *Verticillium chlamydosporium* are responsible for storage rot. 85% of the rhizomes possess the mold growth of *Penicillium brevicompactum* [143].

Pathogenicity test is available for *Acremonium murorum, Acrostalagmus luteoalbus, Fusarium* sp., *F. oxysporum, Lasiodiplodia theobromae* and *Sclerotium rolfsii* associated storage rot [36].

Figure 5.
Rotting of rhizome due to the infestation of F. oxysporum (source: pestnet.org).

4.4.3 Protection

4.4.3.1 Chemical methods

Incidence of storage rots is reduced while the rhizome is treated with benomyl (750 ppm) and/or gibberellic acid (150 ppm) before the storage [144]. Dipping the rhizomes with imazalil or prochloraz (0.8 g a.i/liter) and then storing at 10°C gave good protection against *Botryodiplodia, Aspergillus, Diplodia, Fusarium, Rhizoctonia* and *Pythium* [145]. Mancozeb and carbendazim together treatment on ginger rhizomes protect the rhizomes from the rot [146, 147]. Application of 0.3% Ridomil MZ during the storage condition causes the low infection [107]. Steeping of rhizomes in carbendazim (0.1%) for 60 min before storage also controlled storage rots and reduced the disease incidence from 71.4 to 18.2% [112].

Treating the rhizome with aureofungin (0.02%) and Benomyl (0.2%) before the Storage, control the disease [148]. Immersing the rhizome in carbendazim (0.1%) for 60 min reduce the disease from 71.4 to 18.2% [112].

Sharma et al. [149] reported that mancozeb fungicides compared to carbendazim is best chemical to protect the rhizome for the longer period from the fungus infection. The presence of Mancozeb was observed at 120th days of storage. But the health point of human, carbendazim treated rhizomes is safe than mancozeb treated rhizome Pre-storage treatment of rhizome with Topsin-M and Bavistin (each at 0.2% concentration for 60 min) reduce the appearance of disease on rhizome, increase the weight of rhizome, surface shriveling and sprouting of rhizomes [31].

4.4.3.2 Cultural practices

Storing the rhizomes in lower temperature avoid weight loss, increase the sprouting but with higher risk of infection when comparing with storage at room temperature. Rhizomes packed in PVC film preserve the weight but has high chance of disease [150]. Dipping the rhizomes in the *Allium sativum* extract or immersing in a suspension of *P. fluorescens* and *T. harzianum* (0.5% for 30 min) before storage, reduce disease incidence [151]. Similarly, dipping in garlic extract (20% w/v for 30 min) also reduces the disease incidence [152].

5. Nematodes causing disease in ginger

Meloidogyne sp., *Radopholus similis* and *Pratylenchus* sp. are major nematodes cause significant loss to ginger.

5.1 Root-knot

Disease causes 74% of reduction in rhizome weight. Nematode infections aggregate the fungus and bacterial infection.

5.1.1 Symptoms

Nematode feeds rhizomes, roots and base of the pseudo stems. In the root, it causes the swellings or knots. The symptoms of root knot are very similar to root gall. Irregular round galls and spindle-shaped enlargements appear on the tap and side roots. Diameter of gall is 3.3 cm. Infested plants show stunting, chlorosis and marginal necrosis of leaves. Roots are often stunted and deformed. Roots and rhizomes exhibit galling and rotting. Cortex of the rhizomes becomes lumpy and cracked during the severe infection. During the infection process, female nematode attains the maturity and emerges from the gall by breaking the epidermis of the rhizome which gives corky in appearance for rhizome. Small, circular, water soaked, slightly brown lesions with mature females of the nematode (**Figure 6**) below the epidermis of the rhizomes are quite numerous in severely infected rhizomes. The lesions serve as entry points for bacteria and fungi, invade, extend the injury into other tissues and destroy the rhizomes.

5.1.2 Causative agent

It is caused by the plant parasitic nematode, *Meloidogyne* sp. They are sedentary, endoparasitic and produce gall. Due to this, crop is severely affected without destroying the rhizomes. Nematode does not produce galls in ginger rhizomes as it occurred in other plants but it invades immature tissues. The infective stage of the nematode is the second stage juveniles, which have lightly sclerotized cephalic framework.

5.1.3 Protection

Crop rotation with non-hosts such as graminaceous and a few antagonistic crops for 1 or 2 years reduces the nematode. Crop rotation with groundnut-mustard was effective in reducing the population of *M. incognita*. Decomposition of soil amendments (dry or green crop residues, oil cakes, meals, sawdust, FYM etc.) in the nematode infested field not only reduce the nematode but also alter the physical, chemical and biotic conditions of the soil which improve the plant growth. Application of nematicides like Thimet or Phorate (12–15 kg/ha) followed by light irrigation before the sowing reduce the incidence of nematode. Biocontrol agents like *Paecilomyces lilacinus*, *Pasteuria penetrans* and *Pseudomonas fluorescens* are effective against *M. incognita*.

5.2 Burrowing nematode

5.2.1 Symptoms

Infected plants exhibits stunting, reduced vigor and tillering. Top most leaves become chlorotic with scorched tips. Infected plants show yellow leaves with less

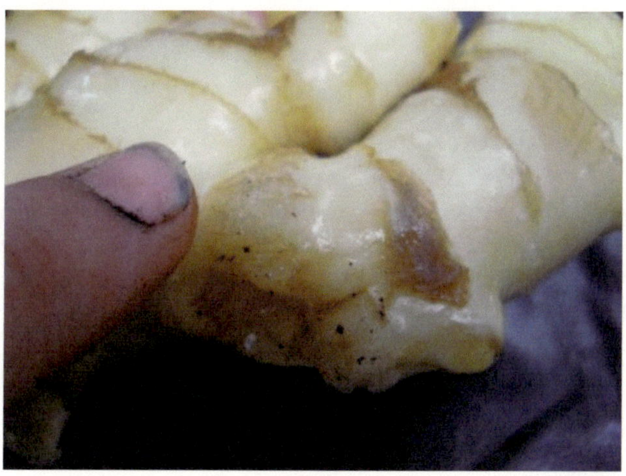

Figure 6.
Brown lesions on the root knot nematode infected rhizome (below) (source: http://www.eastbranchginger.com).

number of shoots and stunted growth. Infected rhizomes possess small, water-soaked shallow lesions which later turn brown. These small lesions merge and rot the rhizome.

5.2.2 Causative agent

The causative agent is Radopholus similis.

5.2.3 Protection

Crop rotation with taro and cassava, applying large qualities of poultry manure, dipping seed in hot water at 51°C for 10 min before sowing, reduce the infection of *Radopholus similis*. Chemical control is not recommended for the control of this disease.

5.3 Lesion nematodes

5.3.1 Symptoms

It affects roots and rhizome severely. It causes extensive damage to cortical tissues of root. Infestation of the nematodes causes yellowing of leaves and dry rot on rhizome. Dark brown necrotic lesion is observed in nematode infected rhizomes. The fingers are severely affected by this nematode.

5.3.2 Causative agent

Several species of *Pratylenchus* namely *P. brachyurus, P. coffeae, P. indicus* and *P. zeae* are infecting ginger.

6. Minor diseases

Some diseases of minor importance have also been reported on ginger like Cercospora leaf spot caused by *Cercospora zingibericola* [153], anthracnose caused by *Colletotrichum zingiberis* [154], *Pyricularia* leaf spot caused by *Pyricularia zingiberi* [155], basal rot caused by *Sclerotium rolfsii* [92, 156] and *Septoria* leaf spot caused by *Septoria zingiberis* [157].

7. Nonparasitic diseases

7.1 Sunburn

Young ginger plants are very susceptible to sunburn when the temperatures exceed 90°F due to high light intensity. Mild sunburn affects only the leaves, but acute sunburn damages the entire shoot [158]. Drought and lack of water may cause the same effects as sunburn.

7.2 Lime-induced chlorosis

Excessive liming or abundant coral sand in the soil may cause yellowing of the blades and poor growth.

8. Disease caused by arthropods

Insect involved in spreading the pathogens responsible for the diseases and also involved in damaging the foliage and rhizomes.

8.1 Shoot borer

8.1.1 Symptoms

The moth lay eggs on the growing bud, petiole or leaf of the young plants. Caterpillars bore through the central shoots, feed the growing buds resulting in withered and dried shoot referred to as "Dead Heart". The presence of a bore hole on the pseudo stem through which frass is extruded and withered and yellow central shoot is a characteristic symptom of pest infestation.

8.1.2 Causative pest

Shoot borer caused by *Dichocoris punctiferalis* (**Figure** 7).

8.1.3 Management

Ginger is protected from the shoot borer by collecting the entire emerged adult, destroying and by installing light trap for adult mass trapping, destroying infested plant and by chemical application of Metarhizium and treating with *Beauveria bassiana*.

8.2 White grub

8.2.1 Symptoms

It feeds the base of the pseudo stem, roots and newly formed rhizomes. Pest infestation leads to yellowing of the leaves. It make large hole in the rhizome and reduce its market value. The entire crop may be lost in severely infested plantations. The adults are dark brown beetles and measures about 2.5 mm x 1.5 mm in size. The grubs are creamy white and live in soil.

Figure 7.
Dichocoris punctiferalis cause shoot borer in the ginger (source: Ref. [159]).

8.2.2 Causative pest

White grub caused by *Holotrichia* spp. (**Figure 8**).

8.2.3 Management

Leaving the land fallow for 2 consecutive years reduce the pest population. Growing of resistant crops such as sunflower also checks the buildup of grub population. Sowing of trap crops such as sorghum, maize and onion reduce the white grub infestation. Application of *Beauveria bassiana* or *Metarhizium anisopliae* with vermicompost (5 g/kg) or drenching the soil with these entomopathogenic fungi (5 g/l) reduces the pest. Two sprays of neem oil 0.15 EC (1500 ppm) at 3 ml/l at 15 days interval is found to be effective.

8.3 Leaf roller

8.3.1 Symptoms

It is an olive green caterpillar with a distinct black head. It folds the leaves and stays inside the fold and defoliates the leaves from the tip and margins. When one portion is complete, it moves and makes another fold.

8.3.2 Causal organism

Leaf roller is caused by *Udaspes folus* (**Figure 9**).

8.3.3 Protection

Maintain the field sanitation and application of *Bacillus thuringiensis* (1–2 g/l) prevents the leaf roller.

8.4 Chinese rose beetle

The Chinese rose beetle, *Adoretus sinicus* (Burm.), (**Figure 10**) is pale reddish-brown and has nocturnal feeding habits. The damage on the foliage is characteristic, being peppered or shot with holes, or more or less skeletonized. The eggs, which are

Figure 8.
Holotrichia responsible for white grub in ginger (source: Ref. [159]).

Figure 9.
Udaspes folus causes leaf roller in ginger (source: Ref. [159]).

Figure 10.
Chinese rose beetle (source: Vikaspedia).

laid in the soil, hatch in about 4 days, and the entire life cycle from egg to adult takes about 6–7 weeks. The larvae apparently feed on decaying plant material; pupation occurs in the soil. During the daytime, adults are usually found hiding in loose soil or among dead leaves. This pest is prevalent in the field at low elevations [160].

8.4.1 Protection

Chinese rose beetles are repelled by bright light and so shining the plants with bright light deter them from feeding; covering young plants with e.g. floating row covers can help to protect plants until they are old enough to withstand attacks by the beetle.

8.5 Fullers rose beetle

The Fullers rose beetle, *Pantomorus godmani* (Crotch), feeds ginger leaves [160]. The adults are 7–9 mm long and more prevalent at higher elevations, apparently requiring cooler temperatures for survival.

8.6 Grasshoppers

Ginger leaves are occasionally damaged by grasshoppers. This occurs at times when there is a high population incidence of these insects [160].

8.7 Scavenger flies

These flies breed in decayed plant tissues of ginger. These include *Eumerus marginatus* (Grims) (syrphid fly), *Euxesta quadrivittata* (Macq.) (otitid scavenger fly) and sciarid gnat fly. The importance of these insects in ginger production remains to be determined. Syrphid fly may be a threat to ginger when the crop becomes dormant. Sometime, the eggs laid on the decaying flower stalk, develop into larva, reaches the rhizome and reduce the quality of rhizome [160].

Author details

Gupta Meenu[1] and Tennyson Jebasingh[2*]

1 Department of Vegetable Science, Dr YS Parmar University of Horticulture and Forestry, Solan, India

2 Department of Plant Sciences, School of Biological Sciences, Madurai Kamaraj University, Madurai, Tamil Nadu, India

*Address all correspondence to: jebasinghs@gmail.com

References

[1] Agricultural and Processed Food Products Export Development Authority (APEDA). 2019

[2] Paret ML, Cabos R, Kratky BA, Alvarez AM. Effect of plant essential oils on *Ralstonia solanacearum* race 4 and bacterial wilt of edible ginger. Plant Disease. 2010;**94**:521-527

[3] Sharma BR, Dutta S, Roy S, Debnath A, Roy MD. The effect of soil physicochemical properties on rhizome rot and wilt disease complex incidence of ginger under hill agro climatic region of West Bengal. Journal of Plant Pathology. 2010;**26**:198-202

[4] So IY. Studies on ginger mosaic virus. Korean Journal of Plant Protection. 1980;**19**(2):67-72

[5] Nambiar KKN, Sarma YR. Mosaic diseases of ginger (*Zingiber officinale*). Arecanut and Spices Bulletin. 1974;**6**:3-4

[6] Thomas JE. Purification and properties of ginger chlorotic fleck virus. The Annals of Applied Biology. 1986;**108**(1):43-50

[7] Hull, R. The grouping of small spherical plant viruses with single RNA components. The Journal of General Virology. 1977;**36**:289-295

[8] Gao SL, Bian YY, Chen BJ. Tissue culture of ginger to control virus diseases and rapid high yielding cultivation. China Vegetables. 1999;**3**:40-41

[9] James CG, Michael J, Melzer J, Sugano SK, Wayne BB, John SH. Viruses in flowering ginger. Plant Disease. 2018;**PD-116**

[10] Kumar A, Sarma YR. Characterization of *Ralstonia solanacearum* causing bacterial wilt of ginger in India. Indian Phytopathology. 2005;**57**:12-17

[11] Dake JN, Ramachandran N, Sarma YR. Strategies to control rhizome rot (*Pythium* spp.) and bacterial wilt (*Pseudomonas solanacearum*) of ginger. Journal of Coffee Research. 1988;**18**:68-72

[12] Pegg KG, Moffett ML, Colbran RC. Disease of ginger in Queensland. Queensland Agricultural Journal. 1974;**100**:611-618

[13] Hayward AC, Moffett ML, Pegg KG. Bacterial wilt of ginger in Queensland. Queensland Journal of Agricultural and Animal Science. 1967;**24**:1-5

[14] Mulya K, Shiomi T, Oniki M. Bacterial wilt disease on industrial crops in Indonesia. Industrial Crops Research Journal. 1990;**2**:30-36

[15] Samuel M, Mathew J. Role and association of root knot nematode *Meloidogyne incognita* in induction of bacterial wilt of ginger incited by *Pseudomonas solanacearum*. Indian Phytopathology. 1986;**36**:398-399

[16] Nelson S. Bacterial Wilt of Edible Ginger in Hawai'i. University of Hawai'i, College of Tropical Agriculture and Human Resources; 2013. 8 p

[17] Paret ML, de Silva AS, Criley RA, Alvarez AM. Detection of Ralstonia solanacearum with an immunostrip assay; its specificity and sensitivity. Indian Phytopathology. 2008;**61**:518-522

[18] Yu Q, Alvarez AM, Moore PH, Zee F, Kim MS, de Silva A, et al. Molecular diversity of *Ralstonia solanacearum* isolated from ginger in Hawaii. Phytopathology. 2003;**93**:1124-1130

[19] Kumar A, Abraham S. PCR based detection of bacterial wilt

pathogen, *Ralstonia solanacearum* in ginger rhizomes and soil collected from bacterial wilt affected field. Journal of Spices Aromatic Crops. 2008;**17**:109-113

[20] Kumar A, Anandaraj M. Method for isolation of soil DNA and PCR based detection of ginger wilt pathogen, *Ralstonia solanacearum*. Indian Phytopathology. 2006;**59**:154-160

[21] Thammakijjawat P, Thaveechai N, Kositratana W, Chunwongse J, Frederick RD, Schaad NW. Detection of *Ralstonia solanacearum* in ginger rhizomes by real-time PCR. Canadian Journal of Plant Pathology. 2006;**28**:391-400

[22] Shan W, Yang X, Ma W, Yang Y, Guo X, Guo J, et al. Draft genome sequence of *Ralstonia solanacearum* Race 4 Biovar 4 strain SD54. Genome Announcement. 2013;**1**(6):e00890-13

[23] Janse J. Potato brown rot in Western Europe—History, present occurrence and some remarks on possible origin, epidemiology and control strategies. Bulletin OEPP/EPPO. 1996;**26**:679-695

[24] Swanson JK, Yao J, Tans-Kersten JK, Allen C. Behavior of *Ralstonia solanacearum* race 3 biovar 2 during latent and active infection of geranium. Phytopathology. 2005;**95**:136-114

[25] Kumar P, Sood AK. An ecofriendly approach for the management of bacterial wilt of tomato. Plant Disease Research (Ludhiana). 2005;**20**:55-57

[26] Tsang MMC, Shintaku M. Hot air treatment for control of bacterial wilt in ginger root. Applied Engineering in Agriculture. 1998;**14**:159-163

[27] Ojha KL, Yadav BP, Bhagat AP. Chemical control of bacterial wilt of ginger. Indian Phytopathology. 1986;**39**:600-601

[28] Sinha SK, Singh VN, Singh DN. Control of bacterial wilt of ginger (*Zingiber officinale*) with antibiotics. Journal of Research, Birsa Agricultural University. 2000;**12**:41-43

[29] Yang W, Xua Q, Liua HX, Wanga YP, Wanga YM, Yangb HT, et al. Evaluation of biological control agents against *Ralstonia* wilt on ginger. Biological Control. 2012;**62**:144-151

[30] Indrasenan G, Kumar KV, Mathew J, Mamen MK. The mode of survival of *Pseudomonas solanacearum* (Smith) Smith causing bacterial wilt of ginger (*Zingiber officinale* Rosc.). Agricultural Research Journal of Kerala. 1981;**19**:93-95

[31] Dohroo NP. Etiology and management of storage rot of ginger in Himachal Pradesh. Indian Phytopathology. 2001;**54**:49-54

[32] Joshi LK, Sharma ND. Diseases of ginger and turmeric. In: Nair MK, Premkumar T, Ravindran PN, Sarma YR, editors. Proc. Nat. Sem. Ginger Turmeric. Kasaragod, Calicut: CPCRI; 1980. pp. 104-119

[33] Butler EJ. An account of genus *Pythium* and some Chytridiaceae. Memoirs of the Department of Agriculture India. 1907;**1**:70

[34] Dohroo NP. Diseases of ginger. In: Ravindran PN, Babu KN, editors. Ginger, the Genus Zingiber. Boca Raton: CRC Press; 2005. pp. 305-340

[35] ISPS. Experiences in collaboration. Ginger pests and diseases. Indo-Swiss Project Sikkim Series 1; 2005. 75 p

[36] Moreira SI, Dutra DC, Rodrigues AC, Oliveira JR, Dhingra OD, Pereira OL. Fungi and bacteria associated with post-harvest rot of ginger rhizomes in Espírito Santo, Brazil. Tropical Plant Pathology. 2013;**38**:218-226

[37] Wang PH, Chung CY, Lin YS, Yeh Y. Use of polymerase chain reaction to detect the soft rot pathogen, Pythium myriotylum, in infected ginger rhizomes. Letters Applied Microbiology. 2003;**36**:116-120

[38] Yella R, Gogoi R, Gogoi G, Phookan AK. A simple technique for producing oospores in *Pythium myriotylum*, causing soft rot of ginger. Indian Phytopathology. 2006;**59**:368-369

[39] Le DP, Smith MK, Aitken EAB. Genetic variation in Pythium myriotylum based on SNP typing and development of a PCR-RFLP detection of isolates recovered from Pythium soft rot ginger. Letters in Applied Microbiology. 2017;**65**(4):319-326

[40] Thomas KM. Detailed Administration Report of the Government Mycologist, Madras, 1937-38. 1938. 21 p

[41] Dake JN. Diseases of ginger (*Zingiber officinale* Rosc.) and their management. Journal of Spices and Aromatic Crops. 1995;**4**:40-48

[42] Bennett MA, Callan NW, Fritz VA. Seed treatments for disease control. Horticultural Science and Technology. 1991;**1**:84-87

[43] Pordesimo AN, Raymundo SA. Rhizome rot of ginger and its control. Coffee and Cocoa Research Journal. 1963;**5**:240

[44] Quimio AJ, Chan HH. Survival of *Pseudomonas solanaeearum* in the rhizosphere of some weed and economic plant species. Philippine Phytopathology. 1979;**15**(1):108-121

[45] Lee WH, Cheong SS, So IY. Properties of suppressive and conducive soils to ginger rhizome rot. Korean Journal of Plant Pathology. 1990;**6**(1):338-342

[46] Hoppe PE. *Pythium* species still viable after 12 years in air-dried muck soil. Phytopathology. 1966;**56**:1411

[47] Dohroo NP, Sharma SL. Evaluation of fungicides for the control of rhizome rot of ginger in storage. Indian Phytopathology. 1986;**36**:691-693

[48] Thakore BBL, Mathur S, Singh RB. Effect of rhizome treatment with fungicides for economic control of rot. Journal of Phytological Research. 1988;**1**:83-84

[49] Singh AK. Management of rhizome rot caused by *Pythium, Fusarium* and *Ralstonia* spp. in ginger (*Zingiber officinale*) under natural field conditions. Indian Journal of Agricultural Sciences. 2011;**81**:268-270

[50] Rajan PP, Gupta SR, Sarma YR, Jackson GVH. Diseases of ginger and their control with *Trichoderma harzianum*. Indian Phytopathology. 2002;**55**:173-177

[51] Smith M, Abbas R. Controlling Pythium and associated pests in ginger. RIRDC Publication No. 11/128. Canberra; 2011

[52] Doshi A, Mathur S. Symptomatology, interaction and management of rhizome rot of ginger. Xenobiotics. 1987;**26**:261-265

[53] Dohroo NP, Sharma SL, Bhardwaj SS. Efficacy of soil applied fungitoxicants against rhizome rot of ginger. Indian Journal of Plant Protection. 1984;**12**:59-60

[54] Rathaiah Y. Control of soft rot of ginger with Ridomil. Pesticides. 1987;**21**:29-30

[55] Ramachandran N, Dake GN, Sarma YR. Effect of systemic fungicides on *in vitro* growth of *Pythium aphanidermatum,* the rhizome

rot pathogen of ginger. Indian Phytopathology. 1989;**42**:463-465

[56] Srivastava LS. Management of soft rot of ginger in Sikkim. Plant Disease Research. 1994;**9**:146-149

[57] Nath PD. Effect of shade and treatment for rhizome rot of ginger. Annals of Agricultural Research. 1993;**14**:327-328

[58] Rathore VRS, Mathur K, Lodha BC. Activity of volatile and non-volatile substances produced by *Trichoderma viride* in ginger rhizome rot pathogen. Indian Phytopathology. 1992;**45**:253-254

[59] Shanmugam V, Gupta S, Dohroo NP. Selection of a compatible biocontrol strain mixture based on cocultivation to control rhizome rot of ginger. Crop Protection. 2013a;**43**:119-127

[60] Bhai RS, Kishore VK, Kumar A, Anandaraj M, Eapen SJ. Screening of rhizobacterial isolates against soft rot disease of ginger (*Zingiber officinale* Rosc.). Journal of Spices and Aromatic Crops. 2005;**14**:130-136

[61] Dohroo NP, Kansal S, Mehta P, Ahluwalia N. Evaluation of eco-friendly disease management practices against soft rot of ginger caused by *Pythium aphanidermatum*. Plant Disease Research. 2012;**27**:1-5

[62] Rakesh KN, Dileep N, Noor Nawaz AS, Junaid S. Antifungal activity of cow urine against fungal pathogens causing rhizome rot of ginger. Environment and Ecology. 2013;**31**:1241-1244

[63] Gupta SL, Paijwar MS, Rizvi G. Biological management of rot disease of ginger (*Zingiber officinale* Rosc.). Trends in Biosciences. 2013;**6**:302

[64] Abbasi PA, Riga E, Conn KL, Lazarovits G. Effect of neem cake soil amendment on reduction of damping-off severity and population densities of plant-parasitic nematodes and soil borne plant pathogens. Canadian Journal of Plant Pathology. 2005;**27**:38-45

[65] Poudyal BK. Jeevatu: One of the best bio-agents for the control of soft rot of ginger. In: 2nd International Conference on Environment Science and Biotechnology IPCBEE. Vol. 48. Singapore: IACSIT Press; 2012. pp. 66-70

[66] Ram D, Mathur K, Lodha BC, Webster J. Evaluation of resident biocontrol agents as seed treatments against ginger rhizome rot. Indian Phytopathology. 2000;**53**:450-454

[67] Shanmugam V, Thakur H, Kaur J, Gupta S, Rajkumar S, Dohroo NP. Genetic diversity of *Fusarium* spp. inciting rhizome rot of ginger and its management by PGPR consortium in the western Himalayas. Biological Control. 2013;**66**:1-7

[68] Gupta M, Dohroo NP, Gangta V, Shanmugam V. Effect of microbial inoculants on rhizome disease and growth parameters of ginger. Indian Phytopathology. 2010;**63**:438-441

[69] Praveen T, Sharma K. Management of "soft rot" of ginger by botanicals. International Journal of Pharmaceutical and Life Sciences. 2014;**5**:3478-3484

[70] Le DP, Smith M, Hudler GW, Aitken E. *Pythium* soft rot of ginger: Detection and identification of the causal pathogens and their control. Crop Protection. 2014;**65**:153-167

[71] Dohroo NP. Final ICAR Report on Multilocational Project on Rhizome Rot of Ginger. Solan: UHF; 1993. 38 p

[72] Harvey P, Lawrence L. Managing Pythium root disease complexes to improve productivity of crop rotations. Outlooks Pest Management. 2008;**19**:127-129

[73] Rames EK, Smith MK, Hamill SD, De Faveri J. Microbial indicators related to yield and disease and changes in soil microbial community structure with ginger farm management practices. Australasian Plant Pathology. 2013;**42**:685-692

[74] Sadanandan AK, Iyer R. Effect of organic amendments on rhizome rot of ginger. Indian Cocoa, Arecanut and Spices Journal. 1986;**9**:94-95

[75] Thakore BBL, Mathur S, Singh RB, Chakravarti BP. Soil amendment with oil cakes in ginger field for rhizome rot control. Korean Journal of Plant Protection. 1987;**26**:267-268

[76] Stirling GR, Smith MK, Smith JP, Stirling AM, Hamill SD. Organic inputs, tillage and rotation practices influence soil health and suppressiveness to soilborne pests and pathogens of ginger. Australasian Plant Pathology. 2012;**41**:99-112

[77] Kadam RV, Jagtap GP, Dey U. Management of rhizome rot (*Pythium aphanidermatum*) in ginger through amendments. Journal of Plant Diseases and Science. 2014;**9**:209-213

[78] Kumar A, Avasthe RK, Borah TR, Lepcha B, Pandey B. Organic mulches affecting yield, quality and diseases of ginger in mid hills of north eastern Himalayas. Indian Journal of Horticulture. 2012;**69**:439-442

[79] Indrasenan G, Paily PV. Studies on the soft rot of ginger (*Zingiber officinale* Rosc.) caused by *Pythium aphanidermatum* (Edson) Fitz. Agricultural Research Journal of Kerala. 1974;**11**:53-56

[80] Setty TAS, Guruprasad TR, Mohan E, Reddy MNN. Susceptibility of ginger cultivars to rhizome rot at west coast conditions. Environment and Ecology. 1995a;**13**:242-244

[81] Senapati AK, Sugata G. Screening of ginger varieties against rhizome rot disease complex in eastern Ghat high land zone of Orissa. Indian Phytopathology. 2005;**58**:437-439

[82] Kavitha PG, Thomas G. Evaluation of Zingiberaceae for resistance to ginger soft rot caused by *Pythium aphanidermatum* (Edson) Fitzp. PGR Newsletter Biodiveristy International. 2008;**152**:54-57

[83] Bhai RS, Sasikumar B, Kumar A. Evaluation of ginger germplasm for resistance to soft rot caused by *Pythium myriotylum*. Indian Phytopathology. 2013;**66**:93-95

[84] Mathur K, Ram D, Poonia J, Lodha BC. Integration of soil solarization and pesticides for management of rhizome rot of ginger. Indian Phytopathology. 2002;**55**:345-347

[85] Lokesh MS, Patil SV, Gurumurthy SB, Palakshappa MG, Anandaraj M. Solarization and antagonistic organisms for management of rhizome rot of ginger in Karnataka. International Journal of Plant Protection. 2012;**5**:195-200

[86] Deadman M, Al Hasani H, Al Sa'di A. Solarization and biofumigation reduce *Pythium aphanidermatum* induced damping-off and enhance vegetative growth of greenhouse cucumber in Oman. Journal of Plant Pathology. 2006;**88**:335-337

[87] Dohroo NP, Gupta M. Effect of bioagents on management of rhizome diseases, plant growth parameters and nematode population in ginger. Agricultural Science Digest. 2014;**34**:41-44

[88] Lalfakawma C, Nath BC, Bora LC, Srivastava S, Singh JP. Integrated disease management of *Zingiber officinale* Rosc. rhizome rot. The Bioscan. 2014;**9**:265-269

[89] Dohroo NP, Kansal S, Ahluwalia N. Studies on eco-farmer-friendly practices for management of soft rot of ginger (*Zingiber officinale*). Indian Phytopathology. 2015;**68**:93-96

[90] Simmonds JH. Rep. Dep. Agric. Queensl. 1953-54. Science Branch, Plant Pathology Section. 1955. pp. 55-56

[91] Trujillo EE. Diseases of Ginger (Zingiber officinale) in Hawaii. CIRCULAR 62. Hawaii Agricultural Experimental Station, University of Hawaii. December 1964

[92] Haware MP, Joshi LK. Basal rot of ginger (*Zingiber officinale)* caused by *Sclerotium rolfsii* from Madhya Pradesh. Indian Phytopathology. 1973;**26**:575-576

[93] Dohroo NP. Further studies on rhizome rot of ginger (*Zingiber officinale* Rosc.) [PhD thesis]. Solan, HP: HPKVV; 1982

[94] Yang KD, Kim HM, Lee WH, So IN. Studies on rhizome rot of ginger caused by *Fusarium oxysporum* f.sp. *zingiberi* and *Pythium zingiberum*. Korean Journal of Plant Pathology. 1988;**4**:271-277

[95] Bhardwaj SS, Gupta PK, Dohroo NP, Shyam KR. An addition to fungi causing rhizome rot of ginger. Plant Disease Research. 1988;**3**:66

[96] Dohroo NP. *Pythium ultimum* on *Zingiber officinale*. Indian Phytopathology. 1987;**40**:275

[97] Sharma SL, Dohroo NP. Efficacy of chemicals in controlling rhizome rot of ginger and turmeric. In: (ICAR), CPCRI, Calicut, 8-9 April, 1980. 65p

[98] Chauhan HL, Patel MH. Etiology of complex rhizome rot of ginger (*Zingiber officinale*) in Gujarat and *in vitro* screening of fungicides against its causal agents. Indian Journal of Agricultural Sciences. 1990;**60**:80-81

[99] Dohroo NP, Sharma SK. Variability of *Fusarium oxysporum* f.sp. *zingiberi*, the incitant of ginger yellows. Indian Phytopathology. 1992b;**45**:247-248

[100] Pappallardo L, Smith MK, Hamill B, Stirling CAM, McKay D. DNA amplification fingerprinting analysis of genetic variation within *Fusarium oxysporum* f.sp. *zingiberi*. Australasian Plant Pathology. 2009;**38**:51-54

[101] Gupta M, Jarial K, Vikram A. Morphological, cultural, pathological and molecular variability among *Fusarium oxysporum* f.sp. *zingiberi* isolates. International Journal of Bio-resource and Stress Management. 2014;5(3):375-380

[102] Dohroo NP. Seed transmission of pre-emergence rot and yellows of ginger. Plant Disease Research. 1989;**4**:73-74

[103] Sharma ND, Jain AC. Studies in biocontrol of *Fusarium oxysporum* f.sp. *zingiberi,* the causal organism of yellows disease of ginger. Indian Phytopathology. 1978;**31**:260-261

[104] Sharma SK, Dohroo NP. Effect of soil hydrothermal regimes on the development of ginger yellows. Indian Journal of Plant Pathology. 1989;7:109-111

[105] Meena, Mathur S. Eco-friendly management of rhizome rot of ginger caused by *Fusarium oxysporum* through chemical and bio-agent. Indian Phytopathology. 2005;**29**:238-246

[106] Singh KA, Gomez AA. Statistical Procedure for Agricultural Research. 2nd ed. Wiley International Science Publication; 2001. pp. 28-192

[107] Singh AK, Singh S, Edison S. Effect of shading on the Phyllosticta leaf spot, sun burn of leaves and yield of ginger. Indian Phytopathology. 2004;**57**:197-199

[108] Stirling MR, Akhter N, Chowdhury SM, Ali M, Ahmed KU. Evaluation of fungicide against *Pythium aphanidermatum* causing rhizome rot of ginger. Journal of Agricultural Science and Technology. 2006;**2**:27-30

[109] Usman MB. Management of *Fusarium* and nemic wilts of ginger by grafting, soil amendment, chemicals and bioagents. Indian Phytopathology. 2006;**23**:255-259

[110] Hasnat M, Rajib B, Hossain MA, Anam MM, Kabir H. Effect of chemicals, bio-agent, plant extract and soil amendments in controlling rhizome rot of ginger. International Journal of Natural and Social Sciences. 2014;**1**:1-11

[111] Sagar SD. Investigations on the etiology, epidemiology and integrated management of rhizome rot complex of ginger and turmeric [PhD thesis]. Dharwad: Department of Plant Pathology, University of Agricultural Sciences; 2006

[112] Sharma SK, Dohroo NP. Postharvest management of rhizome rot (*Fusarium oxysporum* f.sp. *zingiberi* Trujillo) of ginger through chemical and antagonist. Indian Cocoa, Arecanut and Spices Journal. 1991;**14**:150-152

[113] Khatso K, Ao NT. Biocontrol of rhizome rot disease of ginger (*Zingiber officinale* Rosc.). International Journal of Bio-resource and Stress Management. 2013;**4**:317-321

[114] Amreen T, Kumar VBS. Sensitivity of *Fusarium oxysporum* f.sp. *zingiberi* causing ginger yellows against antagonist and fungicides. Environment and Ecology. 2013;**31**:663-666

[115] Manasa M, Kambar Y, Pallavi S, Vivek MN, Onkarappa R, Prashith Kekuda TR. Biocontrol potential of *Streptomyces* species against *Fusarium oxysporum* f.sp. *zingiberi* (causal agent of rhizome rot of ginger). Advances in Science and Research. 2013;**4**:1-3

[116] Sagar SD, Kulkarni S, Hegde YR. Management of rhizome rot of ginger by botanicals. International Journal of Plant Sciences. 2007;**2**:155-158

[117] Ramteke PK, Kamble SS. Evaluation of phytoextracts against *Fusarium solani* (Mart.) Sacc. causing rhizome rot of ginger (*Zingiber officinale* Rosc.). Current Biotica. 2011;**4**:469-474

[118] Athawale V, Paralikar P, Ingle AP, Rai M. Biogenically engineered nanoparticles inhibit *Fusarium oxysporum* causing soft-rot of ginger. IET Nanobiotechnology. 2018;**12**(8):1084-1089

[119] Rana KS. Effect of seed selection in the management of yellows disease of ginger. Indian Journal of Mycology and Plant Pathology. 1991;**21**:183-185

[120] Sharma BR, Dutta S, Ray S, Roy S. Influence of plant spacing, seed rhizome size and cultivars on the incidence of rhizome rot and wilt disease complex of ginger. Journal of Horticulture and Forestry. 2012;**4**:105-107

[121] Dohroo NP. Peroxidase and polyphenol oxidase activities in rhizome rot of ginger. Indian Phytopathology. 1989;**42**:167

[122] Rana KS, Arya PS. Rhizome rot and yellows disease of ginger in HP. Indian Journal of Mycology and Plant Pathology. 1991;**21**:60-62

[123] Priya R, Swetha, Subramanian RB. Isolation and molecular analysis of R-gene in resistant *Zingiber officinale* (ginger) varieties against *Fusarium oxysporum* f.sp. *zingiberi*. Bioresource Technology. 2008;**99**:4540-4543

[124] Dohroo NP. Integrated management of yellows of ginger. Indian Phytopathology. 1995;**48**:90-92

[125] Ramakraishanan TS. A leaf spot disease of *Zingiber officinale* caused by *Phyllosticta zingiberi* n.sp. Proceedings of the Indian Academy of Sciences: Section B. 1942;**20**:167-171

[126] Brahma RN, Nambiar KKN. Survival of Phyllosticta zingiberi Ramakr., causal agent of leaf spot of ginger. In: Nair MK, Prem Kumar T, Ravindran PN, Sarma YR, editors. Proceedings of National Seminar on Ginger and Turmeric; CPCRI, Kasargod. 1982. pp. 123-125.

[127] Cerezine PC, Olinisky IA, Bittencourt MVL, Valerio Folho WV. *Phyllosticta* leaf spot on ginger. Cultural characterization of the pathogen and effect of chemical treatment on disease control in Morrestes, Parana state, Brazil. Pesquisa Agropecuaria Brasileira. 1995;**30**:477-487

[128] Brahma RN, Nambiar KKN. Spore release and dispersal in ginger leaf spot pathogen *Phyllosticta zingiberi*. In: Bavappa KVA et al., editors. Proceedings of National PLACROSYM-V 1982. Placrosym Standing Committee. 1984

[129] Sood R, Dohroo NP. Epidemiology and management of leaf spot of ginger in Himachal Pradesh. Indian Phytopathology. 2005;**58**:282-288

[130] Senapati AK, Mukharjee AK, Ghose S. Identification of resistance sources in ginger cultivars against Phyllosticta leaf spot. Indian Journal of Plant Protection. 2012;**40**:80-81

[131] Singh AK. Efficacy of fungicides for the control of leaf spot disease of ginger under the field conditions of Chhattisgarh (India). African Journal of Agricultural Research. 2015;**10**:1301-1305

[132] Sohi HS, Sharma SL, Verma BR. Chemical control of *Phyllosticta* leaf spot of ginger (*Zingiber officinale*). Pesticides. 1973;7:21-22

[133] Grech NM, Frean RT. Ginger leaf spot: A cause for concern in the ginger industry. Citrus Grow. Subtrop. Fruit J. 1988;**644**:14-15

[134] Verma RK, Vyas SC. Persistence and protective activity of some fungicides in relation to Phytllosticta leaf spot of ginger. Indian Journal of Mycology and Plant Pathology. 1981;**11**:14-16

[135] Patiram Upadhyaya RC, Singh LN. An appraisal of ginger (*Zingiber officinale* Rosc.) production in Sikkim. Journal of Spices and Aromatic Crops. 1995;**4**:111-118

[136] Setty TAS, Guruprasad TR, Mohan E, Reddy MNN. Susceptibility of ginger cultivars to Phyllosticta leaf spot at west coast conditions. Environment and Ecology. 1995b;**13**:443-444

[137] Dohroo NP, Shyam KR, Bhardwaj SS, Korla BN. Reaction of ginger germplasm to Phyllosticta leaf spot. Indian Phytopathology. 1986;**39**:650-606

[138] Rao TNG, Sasikumar B, George JK. Field reaction of ginger germplasm to *Phyllosticta zingiberi*. Indian Phytopathology. 1995;**48**:463-465

[139] Sharma ND, Jain AC. A checklist and selected bibliography of ginger diseases of the world. PANS. 1977;**23**:474-481

[140] Mishra B, Rath GC. *Geotrichum* rot of stored ginger. Indian Journal of Mycology and Plant Pathology. 1989;**18**:213

[141] Geeta GS, Reddy TKR. *Aspergillus flavus* link and its occurrence in relation to other mycoflora on stored spices. Journal of Stored Products Research. 1990;**26**:211-213

[142] Dohroo NP, Sharma M. New host records of fungi from India. Indian Phytopathology. 1992a;**45**:280

[143] Overy DP, Frisvad JC. Mycotoxin production and postharvest storage rot of ginger (*Zingiber officinale*) by *Penicillium brevicompactum*. Journal of Food Protection. 2005;**68**:607-609

[144] Okwouwulu PA, Nnodu EC. Some effect of pre-storage chemical treatments and age at harvesting on the storability of fresh ginger rhizomes (*Zingiber officinale* Rosc.). Tropical Science. 1988;**28**:123-125

[145] Grech NM, Swarts DH. Postharvest application of fungicides for control of fungal decay of ginger rhizomes stored under stimulated low temperature shipping conditions. Phytophylactica. 1990;**22**:457-458

[146] Dohroo NP, Bhardwaj SS, Shyam KR. Management of rhizome rot of ginger in storage through fungitoxicants. Pestology. 1986;**10**:24-25

[147] Dohroo NP, Malhotra R. Control of storage rot of ginger in Himachal Pradesh. In: Gupta VK, Sharma RC, editors. Integrated Disease Management and Plant Health. Jodhpur: Scientific Publishers; 1995. pp. 199-202

[148] Haware MP, Joshi LK, Sharma ND. Effect of postharvest treatment of aureofungin on rhizome rot and viability of ginger seed rhizomes. Hindustan Antibiotics Bulletin. 1973;**15**:84-85

[149] Sharma ID, Dohroo NP, Dubey JK, Korla BN. Monitoring mancozeb and carbendazim residues in ginger (*Zingiber officinale* Rosc.) following postharvest dip. Plant Disease Research. 1992;7:13-15

[150] Lana MM, Casali VWD, Finger FL, Reis FP. Evaluation of postharvest storage of ginger rhizomes. Horticultura Brasileira. 1993;**11**:139-141

[151] Ram J, Thakore BBL. Management of storage rot of ginger by using plant extracts and biocontrol agents. Journal of Mycology and Plant Pathology. 2009;**39**:475-479

[152] Jadhav SN, Aparadh VT, Bhoite AS. Plant extract using for management of storage rot of ginger in Satara Tehsil (M.S.). International Journal of Pharmaceutical and Phytopharmacological Research. 2013;**4**:1-2

[153] Kar AK, Mandal M. New *Cercospora* spp. from West Bengal. Transactions of the British Mycological Society. 1969;**53**:337-360

[154] Nema KG, Aggarwal GP. Fungi causing plant diseases at Jabalpur (Madhya Pradesh-IV). Proceedings of the National Academy of Sciences, India Section B. 1960;**30**:55-58

[155] Rathaiah Y. *Pyricularia* leaf spot of ginger in Assam. Indian Phytopathology. 1979;**32**:321-322

[156] Mehrotra BS. *Fusarium roseum* and *Sclerotium rolfsii* on ginger rhizomes. Indian Phytopathology. 1952;**5**:52-54

[157] Sundaram NV. Notes on some fungi from South India. Indian Phytopathology. 1961;**14**:202-209

[158] Grossman HM. Ginger production. Queensland Agricultural Journal. 1954;**78**:259-262

[159] Integrated Pest and Disease Management in Ginger by IPM/State Bio-Control Lab. Gangtok, East Sikkim: Food Security & Agriculture Development Department, Horticulture & Cash Crops Development Department Tadong

[160] Trujillo EE. Diseases of ginger (*Zingiber officinale*) in Hawaii. In: CIRCULAR 62. Hawaii Agricultural Experimental Station, University of Hawaii; 1964

Section 3

Antimicrobial Potentials

Chapter 6

Harnessing the Therapeutic Properties of Ginger (*Zingiber officinale* Roscoe) for the Management of Plant Diseases

Elias Nortaa Kunedeb Sowley and Frederick Kankam

Abstract

Ginger (*Zingiber officinale* Roscoe) is one of the most widely used spices in the world. The therapeutic benefits of ginger are mainly due to the presence of volatile oils, phenols, alkaloid, and high oleoresin content. Ginger extracts have been extensively studied for a broad range of biological activities including antibacterial, antifungal, antiviral, anticonvulsant, analgesic, antiulcer, gastric antisecretory, and antitumor. This is all the more necessary because ginger is of plant origin, specifically more biodegradable, readily available, cheaper, and environmentally friendlier than synthetic chemicals. Since, some farmers in developing countries use ginger extracts as traditional medicine in the treatment of human diseases, it will be easy for them to adopt these extracts as biopesticides for the management of plant diseases. This book chapter seeks to outline the bioactive compounds and therapeutic benefits of ginger in plant disease management, and the mechanisms of action are also discussed.

Keywords: antibacterial, antifungal, biodegradable, synthetic chemicals, ginger

1. Introduction

Ginger (*Zingiber officinale* Roscoe) is a spicy aromatic plant from the family Zingiberaceae. There are about 150 species in the genus *Zingiber*, but *Zingiber officinale* is the widely cultivated and spicy species [1]. Ginger is largely grown in the tropics [2]. In China, ginger has been useful in various indigenous medicines over the centuries [3]. Ginger is regarded as a general medicinal material in the Chinese ayurvedic culture for the treatment of digestion-related discomforts [4]. According to the Transparency Market Research [5] report, ginger is among the high-valued and economic herbal commodities of about the 6.5% per year projected increase in market value which could go up as high as US$ 4.18 billion with about 7.5% estimated rapid growth in consumption by 2022.

Ginger has several uses or functions in our daily lives. It is importantly used in households, pharmaceutical, brewery, food, and other related industries to manufacture products such as ginger oil, ginger wine, gingerbread, ginger cake, ginger spice, ginger syrup, ginger drink, and ginger coffee [6]. Ginger oil is produced in economic and commercial quantities in countries like Australia, China,

Indonesia, and India. Recently, the oil obtained from ginger has been found to protect and maintain the kidney against toxicity [7]. Ginger is a medicinal plant and antimicrobial agent. It contains gingerols as its primary bioactive compound with high flavonoid, phytochemical, and pharmacological effects [3]. Various studies in vivo, in vitro, and clinical analysis have over the years been carried out and, thus, affirmed ginger's therapeutic properties which cannot be downplayed. For instance, some of the volatile oils contained in red ginger, namely, trimethyl-heptadien-ol, ar-curcumene, camphene, carbaldehyde, sesquiphellandrene, and nerol, were found to inhibit the growth of bacteria including *Bacillus cereus*, *Escherichia coli*, *Salmonella typhimurium*, and *Pseudomonas aeruginosa* [8].

2. Nutritional components of ginger

Ginger is extensively used in various traditional and manufactured foods as a result of its richness in essential nutrients. The rhizome, which is the principal economic part of the ginger plant, possesses good amount of carbohydrate, minerals, and vitamins, among others. Ginger rhizome is a rich source of minerals including iron, calcium, and phosphorous. It also contains vitamins such as thiamine, riboflavin, niacin, and vitamin C. Ginger rhizomes also possess a potent proteolytic enzyme called zingibain [9]. Torch ginger (*Etlingera elatior* Jack.) inflorescence contains high amounts of dietary fiber, unsaturated fatty acids (palmitoleic acid, linoleic acid, and oleic acid), and essential amino acids (leucine and lysine) [10]. The inflorescence of torch ginger is enriched with essential minerals such as K (1589 mg/100 g), Ca (775 mg/100 g), Mg (327 mg/100 g), P (286 mg/100 g), and S (167 mg/100 g) with lower levels of heavy metal contaminants (Cd, As, Pb, Hg, Ni) [11, 12]. Raw ginger is also reported to contain useful minerals like Mg, Ca, Mn, Fe, Cu, and Zn [13]. Studies on ginger rhizomes obtained from Malaysia and Nigeria showed higher moisture (90.9% vs. 76.9%), crude fiber (3.8 g/100 g), and lower carbohydrate content (6.3 g/100 g sample) than the USDA database [14]. In other studies, inductively coupled plasma-mass spectrometry-based multi-elemental profiling was used to evaluate the quantitative complement of elements and nutritional quality of ginger rhizome, and the result revealed an abundance of 18 elements quantified [13]. The acid-digested rhizomes were found to have K > Mg > Fe > Ca > Na > Mn > Zn > Ba > Cu > Cr > Ni > Pb > Co > Se > As > Be > Cd metals in that order of abundance. Generally, it is supposed that paradol, formed on hydrogenation of shogaol, in ginger plant contains significant antioxidant content which produces protective health benefits in various diseases [15].

3. Bioactive components of ginger

Ginger is a rich source of some important bioactive molecules and compounds such as phenolic groups, alkaloids, and steroids which have medicinal effect [14]. The main aromatic agent of the rhizome is the zingiberol [3]. It has embedded in it some bioactive compounds such as shogaols, zingerone, paradols, and gingerols, which are structurally shown (**Figure 1**). These components do not easily vaporize and are responsible for the "burning" effect, felt in the mouth [16]. In addition to the main bio-compounds, ginger also contains other sub-compounds like 4-gingerol, 6-gingerol, 8-gingerol, 10-gingerols, 6-shogaols, 14-shogaols (**Figure 1**), and many other identified components which are reactive against inflammation [17] (**Table 1**).

The most identified and investigated components of ginger which act against oxidation reactions in biological systems are shogaols and gingerols, possessing a lot

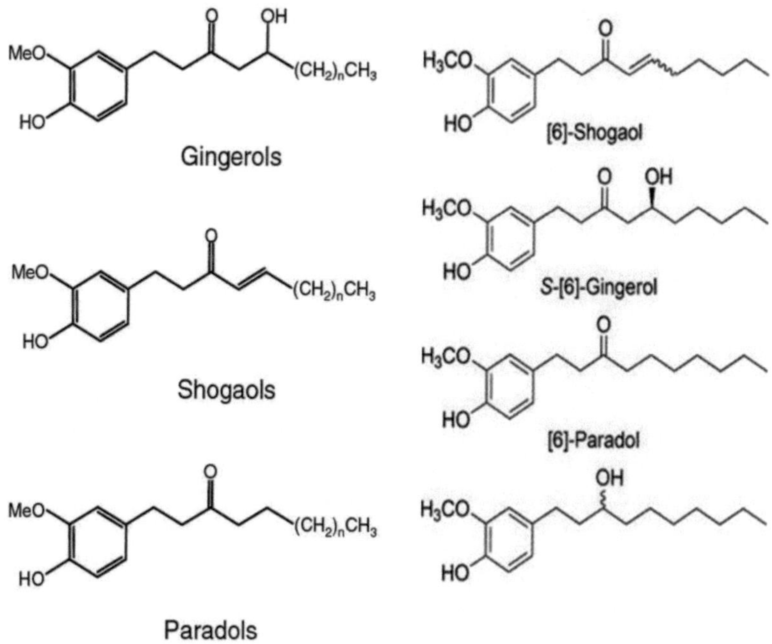

Figure 1.
Chemical structure of major bioactive compounds of ginger.

Ginger constituent	Mechanism of actions	References
Zingerone	Act against inflammation of cells Moderation of prostaglandins synthesis	[19]
10-Gingerol	Inhibition of cell expression, declining the rate of cell multiplication, halt and weakening of the S phase cell cycle	[20]
Paradol	Controlling blood sugar levels Enhances cells recovery and resistance against cancer Promoting the formation and action of antibodies	[21]
8-Gingerol	Inhibition of cancer cell growth Antiplatelet actions Inhibition of blood vessels generations	[22]
Flavonoid	Antioxidant actions against reactive oxygen, which can endanger plant and cells. Protecting cells against risks of damage by oxidation related imbalances	[23, 24]
Oleoresin	Antimicrobial actions against the growth of plants and animal disease microorganism like *Staphylococcus aureus*, *Candida albicans*, *Escherichia coli*	[25–27]
6-Gingerol	Causing cell cycle actions in the G0/G1-phase to cease Reducing cyclin E1, cyclin A, and cyclin D1 levels. Increasing caspase manifestation Stopping the sensing channel of mammalian target of rapamycin (mTOR)	[28]
Phenolic acid	Inhibiting reactive nitrogen and oxygen generation in plants and animal cells, as a result of its ability to stop certain enzymes from action Removing reactive oxygen and nitrogen from cells	[24, 29]
6-Shogaol	Suppress prostaglandin E2 and nitric oxide generation	[30]
Essential oil	Inhibitory actions against microbes Antioxidative reactions	[25, 27]

Table 1.
Some ginger constituents and possible means of bio-actions.

of therapeutic properties like anti-hepatotoxicity and antiprostaglandin production among others [18]. Ginger contains some chemical derivatives and other constituents aside the primary bioactive compounds, which are also highly reactive and with useful therapeutic mechanisms (**Table 1**).

4. Methods of extraction of components of ginger

Before the introduction of modern methods, conventional methods such as infusion, decoction, and percolation, which are a direct simple solvent extraction, were used [31]. Although new techniques have been developed, these conventional methods are still used in Phytochemistry Laboratories. Traditional extraction techniques, though still in use, pose several challenges during extraction. These methods consume a lot of energy and time as well as require large amount of solvents and are also difficult to automate. In recent times, modern extraction techniques such as microwave-assisted, ultrasound-assisted extraction, negative pressure cavitation extraction, and accelerated solvent extraction have been developed to curb the challenges associated with the traditional extraction methods. These techniques are easy to automate and require shorter extraction time and smaller amount of solvents [31].

5. Pharmacological properties

Ginger is used in herbal treatment for colds and other viral infections, poor appetite, digestive problems, arthritis, and headache [32]. Ginger and its constituents have antiemetic, antithrombotic, anti-inflammatory, and antioxidant effects [33]. The major pharmacological activity of ginger appears to be due to gingerol and shogaol [34]. Studies have shown that ginger exhibits several pharmacological activities, such as antioxidant, cytotoxic agent, gastrointestinal, cardiovascular disorders and anti-vomiting, anti-inflammatory, antimicrobial action, and pesticidal effects [34].

5.1 Antioxidant effects of ginger

Many cell culture studies have revealed the antioxidant actions of ginger [35–38]. Some progressive biological studies also show the protective actions of gingerol and ginger extract on many tissues against shocks on the account of several actions that cause oxidation [35]. Ginger is a strong antioxidant substance and may either mitigate or prevent generation of free radicals. Ginger, which is the underground stem or rhizome of the plant *Zingiber officinale Roscoe*, contains polyphenol compounds (6-gingerol and its derivatives), which have a high antioxidant activity [39].

Beverage products obtained through lactic anaerobic biochemical processing of plants from the ginger family contain antioxidant properties [40]. Ginger is one of the prominent herbs noted for its "shock-absorber" actions, which helps to promote insulation for the human system including pregnancy [41, 42]. The antioxidant property of ginger was also confirmed in a study where the extract effectively inhibited rancidity of fats and prevented linoleic acid from oxidizing [43]. The application of ginger also interfered with the generation of nitric oxide [44].

5.2 Cytotoxic properties

The potency of ginger against various forms of cancer like cervical, breast, and prostate have been largely studied [45, 46], and ginger has been found to be effective against the rapid degeneration of cancers, the death of affected cells [22, 47]. Ginger terpenoids were found to promote the stimulation of p53 and thus making the cancer cells relating to endometrium inactive [48]. Rapid multiplication of PC-3 prostate cancer cell was cooperatively prevented by the double mixture of 6-shogaol, 8-gingerol, 6-gingerol, and 10-gingerol [49]. Cell culture studies have demonstrated that ginger served as an inhibitory substance toward cancer of the skin and abnormal growth in the bladder and lung [50, 51].

Recent studies have shown that the administration of ginger powder at two grams per day for a period of 4 weeks suppresses the development of cyclooxygenase-1, an inflammatory-related enzyme which is closely associated with cancer of the large intestine [52]. In similar studies, it was reported that 6-gingerol, a bioactive component responsible for the stingy sensation of ginger, was highly effective against the formation and generation of new blood vessels either within an organism or in an artificial medium outside an organism, and this action could best prevent an abnormal growth and spread of cancers from one cell to the other [53]. Research has revealed that 6-shogaol exhibited inhibitory action against cancer of the breast by stopping the colonization of cells and decreasing of metalloproteinase-9 manifestation [54].

5.3 Gastrointestinal properties of ginger

Ginger has been helpful in the alleviation of numerous gastrointestinal discomforts such as ulcers of the duodenum and other digestive canal. The stomachic mucous membrane is being secured against a number of potential ulcer factors due to its antagonistic reactions toward oxidation [55, 56]. Essence from ginger reportedly blocked *Helicobacter pylori* in a study conducted in an artificial environment [57]. A study has shown that signs of vomiting and nausea were effectively relieved when the powdered form of ginger was administered by mouth, 4 times per day at 250 mg, to 27 gravida women with pregnancy-related morning discomfort for more than a 4-day period [58]. Ginger helps to get rid of constipation when used in the fresh form [59]. The phenolic content in aqueous extract of ginger is reported to have potential ulcer-preventing ability; the aqueous extract of ginger will also reduce free radical damage during ulceration. Hence, ginger is used as an ulcer-preventive agent [60].

5.4 Cardio and antivomitory actions of ginger

Several studies have identified the medicinal properties of shogaol and gingerol bioactive groups of ginger components. Ginger helps to stimulate the muscles of the heart to facilitate blood flow, lower the concentration of blood, and boost metabolic reactions in the cells, which greatly secure the organ systems against offensive muscular contractions [61]. Powdered ginger root in the dose used was found to be effective in reducing nausea and vomiting induced by low-dose cyclophosphamide in combination with drugs causing mild emesis [62].

The 6-gingerol, 6-shogaol, and other ginger-based compounds with two carbonyl groups are catalytically active against the bio-generation of leukotriene, prostaglandin, and thromboxane [63]. Chemo-constituents of ginger like alkaloids, flavonoids, saponins, peptides, and non-primary amino acids showed characteristic actions of dilation and blood pressure reduction [64].

Ginger is a good bio-inhibitory agent against neurotransmitter receptors, which promote flexibility and movement in the lumen of the intestines. Ginger also induces a supportive counteraction in the alimentary canal toward the 5-hydroxytryptamine receptors [55, 56]. Powdered ginger administered at 1 g or more per day suppressed intense and excessive vomiting during pregnancy for about 92% of the studies carried out [65].

5.5 Anti-inflammatory action of ginger

The expression of an inflammation-producing related gene was inhibited in LPS-activated BV2 neuron-supportive cells of the immune system, thereby suppressing neuron-related inflammations [66]. Oral administrations of ginger oil and eugenol against acute arthritis in rats significantly inhibited the expression of joints and paw swollen [67]. The infusion of ginger oil can inhibit prolonged swelling of joints [68].

Ginger structural components like shogaol and gingerol effectively stopped leukotrienes and prostaglandins bio-generations by preventing either 5-lipoxygenase or prostaglandin synthase from expressing [69, 70]. In a published report, ginger was found to have given an outstanding performance against arthritis actions in humans, when consumed in a fresh a state [71]. The study also found out that the generation of many arthritics or inflammation reactions related genes was actively suppressed by using *Alpinia galanga* and essence of ginger [19].

5.6 Antimicrobial action of ginger

Ginger extract has showed antimicrobial activity against a broad spectrum of pathogenic microorganisms. Ginger extract at the rate of 10% has been reported to possess some economic level of anti-pathogenic properties against disease-causing organisms [72]. The floral part and root of ginger contain an extractable oil that averagely act positively against *Bacillus licheniformis*, *Staphylococcus aureus* (gram-positive bacteria), *Klebsiella pneumonia*, and *Pseudomonas stutzeri* (gram-negative bacteria) [73].

An in vitro analysis demonstrated the suppressive ability of ginger essence and other components against the growth and development of contagious bacteria like *Listeria monocytogenes* [74]. Studies have shown that ginger impedes the reproduction and growth of the colon bacteria, some strains of *Salmonella*, carcinogen (23, 24), and *Aspergillus*. Again, the undiluted juice obtained from ginger rhizome has proven to be effective, under room temperature at the rate of 12 and 4% against the development of *Mycoderma* spp. and *Aspergillus niger*, respectively [75].

A study conducted by Chakotiya et al. [76] also revealed how the development of a breed of *Pseudomonas aeruginosa* was suppressed by ginger and its bioactive compounds through the prevention of the synthesis of biofilms. Paste prepared from ginger also exhibited a positive counteraction against the growth of O157:H7 strain of *Escherichia coli* (*E. coli*) when studied in laboratory culture, using beef and laboratory broth [77, 78]. In our previous studies, it was found that the growth of *Aspergillus niger* on yam tubers was suppressed by 65.5% when treated with an ethanol ginger extract [79]. Similarly, *Z. officinale* inhibited the growth of *Rhizopus stolonifer*, *Aspergillus niger*, *Aspergillus flavus*, *Fusarium oxysporum*, and *Botryodiplodia theobromae* significantly [80].

Several reports have demonstrated the effectiveness of ginger against viruses, some parasites, and a group of fungi [81–83]. The highly valued oil extracted from ginger has showed high potency against *Aspergillus flavus* development and the multiplication of certain genes linked to aflatoxin [84, 85]. An ethanol-containing extract obtained from ginger was found to have high efficacy against the multiplication of candida bacteria [86]. Ginger was used to cure tuber rot disease, a fungal disease of an economic importance in yam, by suppressing the growth of *Fusarium oxysporum*, *Penicillium oxalicum*, and *Trichoderma viride,* which are the primary causative agents [87].

6. Pesticidal properties of ginger

It has been reported that ginger extract at 3% exhibited pesticidal potential thereby reducing cabbage looper (*Trichoplusia binotalis*) [88]. Several studies have shown that the methanol extract obtained from ginger can be used in controlling trypanosomiasis, commonly known as "sleeping sickness" due to its counteractions against parasites [89–91]. A good result was achieved when an extract of ginger was evaluated for pestilence against leaf hoppers and defoliators of cowpea (*Vigna unguiculata* L.) in a field study [92].

Studies have also shown that ginger is effective in managing pests like hoppers, root-knot nematodes, aphids, American bollworm, thrips, and mango anthracnose, among others [93]. In a study under both field and laboratory environments, the residues taken from the water extraction process of ginger was found to be active at the rate of about 25–30% in suppressing the oviposition process and development of matured flea beetle of okra (*Callosobruchus maculatus* F.) on the field and in storage [94].

7. Other useful properties of ginger

It has been reported that an ethanolic essence of ginger ultimately reduces the sugar level of blood when given orally to a rat suffering from diabetes. Oral intake of 100 mg/kg of ginger extract, dissolved in 80% of alcohol, gave 38% suppression of yeast-associated fever among rats [95, 96]. 6-Gingerone helps to control obesity by inhibiting and reducing of fats buildup and weight gain among mice [97]. 10-Gingerol exhibited an excellent performance against the growth of *Angiostrongylus cantonensis* larvae, a type of nematode of an economic importance in some regions of the Pacific and Asia [98]. Several studies in recent times have identified ginger as a suppressive medicinal material, which influences recollection ability of the minds and is capable of relieving or preventing neuron inflammations that may help to control infections associated with debasement of neurons [99, 100]. Ginger plays a preventive role against rapid multiplication of cell and growth of cancer-affected cells [101, 102]. Ginger has also been reported to have a good antagonistic action against the virulence of the hepatitis C virus [61, 103].

8. Conclusions

This particular review tried to bring together various research findings and reports of the past, on the therapeutic characteristic properties of ginger. The review also comments on the numerous applications of ginger in the traditional medicine and its pharmaceutical importance. A good number of therapeutic reactions of ginger have been identified, which include anti-inflammatory actions, anti-cancer properties, antioxidant actions, anti-pesticidal properties, cytotoxicity, gastrointestinal actions, antimicrobial effects, and other conventional uses of ginger in plants, animals, and human health. Ginger is a recognized plant in the world of medical and health sciences. It contains gingerols as its primary bioactive compound with high flavonoid and phytochemical and pharmacological effects. Several studies in vivo, in vitro, and clinical analysis have over the years been conducted out and, thus, affirmed ginger's therapeutic properties which cannot be overlook.

Conflict of interest

The authors declare no conflict of interest.

Author details

Elias Nortaa Kunedeb Sowley* and Frederick Kankam
Department of Agronomy, Faculty of Agriculture, University for Development Studies, Tamale, Ghana

*Address all correspondence to: esowley@gmail.com

References

[1] Ravindran PN, Nirmal Babu K. Ginger, the Genus Zingiber. Kerala, India: CRC Press; 2005

[2] Haniadka R, Saldanha E, Sunita V, Palatty PL, Fayad R, Baliga MSA. Review of the gastroprotective effects of ginger (*Zingiber officinale* Roscoe). Food and Function. 2013;**4**(6):845-855. DOI: 10.1039/c3fo30337c

[3] Ali BH, Blunde G, Tanira MO, Nemmar A. Some phytochemical, pharmacological and toxicological properties of ginger (*Zingiber officinale* Roscoe): A review of recent research. Food and Chemical Toxicology. 2008;**46**:409-420. DOI: 10.1016/j.fct.2007.09.085

[4] Tapsell LC, Hemphill I, Cobiac L, Patch CS, Sullivan DR, Fenech M, et al. Health benefits of herbs and spices: The past, the present, the future. Medical Journal of Australia. 2006;**185**:4-24

[5] Transparency Market Research. Ginger Market (Form—Fresh, Dried, Pickled, Preserved, Crystallized, and Powdered; Distribution Channel—Modern Grocery Retail, Traditional Grocery Retail, and Non-Grocery Retail; Application—Culinary, Soups and Sauces, Snacks & Convenience Food, Bakery Products, Alcoholic Beverages, Non-Alcoholic Beverages, and Chocolate and Confectionery)—Global Industry Analysis, Size, Share, Growth, Trends and Forecast 2017—2022. Available online: https://www.transparencymarketresearch.com/ginger-market.html [Accessed: 11/10/2019]

[6] Bijaya BB. Ginger processing in India (*Zingiber officinale*): A review. International Journal of Current Microbiology and Applied Sciences. 2018;7(4):1639-1651. DOI: 10.20546/ijcmas.2018.704.185

[7] Akinyemi AJ, Faboya OL, Paul AA, Olayide I, Faboya OA, Oluwasola TA. Nephroprotective effect of essential oils from ginger (*Zingiber officinale*) and turmeric (*Curcuma longa*) rhizomes against cadmium-induced nephrotoxicity in rats. Journal of Oleo Science. 2018;**67**(10):1339-1345. DOI: 10.5650/jos. ess18115

[8] Rialita T, Rahayu WP, Nuraida L, Nurtama B. Aktivitas antimikroba minyak esensial jahe merah (*Zingiber officinale* var. Rubrum) dan lengkuas merah (*Alpinia purpurata* K. Schum) terhadap bakteri pathogen dan perusak pangan. Ağrı. 2015;**35**:43-52

[9] Agrahari P. A brief study on *Zingiber officinale*, a review. Journal of Drug Discoveries and Therapeutics. 2015;**3**:28. DOI: 10.1016/j.ddstr.2013.05.001

[10] Jeevani Osadee Wijekoon MM, Karim AA, Bhat R. Evaluation of nutritional quality of torch ginger (*Etlingera elatior* Jack.) inflorescence. International Food Research Journal. 2011;**18**(4):1415-1420

[11] Pandotra P, Gupta AP, Khan S, Ram G, Gupta S. A comparative assessment of ISSR, RAPD, IRAP, & REMAP molecular markers in *Zingiber officinale* germplasm characterization. Scientia Horticulturae. 2015;**194**:201-207

[12] World Health Organization (WHO). WHO Traditional Medicine Strategy 2002-2005. Available at: <http://www.wpro.who.int/healthtechnology/bookwho traditionalmedicinestrategy.2000; 2002-2005.pdf> [Accessed: 14/04/2014]

[13] Gupta S, Pandotra P, Gupta AP, Dhar JK, Sharma G, Ram G, et al. Volatile (As and Hg) and non-volatile (Pb and Cd) toxic heavy metals analysis in rhizome of *Zingiber officinale* collected from different

locations of North Western Himalayas by atomic absorption spectroscopy. Food Chemistry and Toxicology. 2010;**48**(10):2966-2971

[14] United States Department of Agriculture. National Nutrient Database for Standard Reference Release 26 Full Report (All Nutrients). Nutrient data for Spices. Ginger. 2013

[15] Badreldin HA, Blunden G, Tanira MO, Nemmar A. Some phytochemical, pharmacological and toxicological properties of ginger (*Zingiber officinale* Roscoe): A review of recent research. Food Chemistry and Toxicology. 2008;**46**:409-420. DOI: 10.1080/10942912.2015.1084633

[16] Dhanik J, Arya N, Nand VA. Review on *Zingiber officinale*. Journal of Pharmacognosy and Phytochemistry. 2017;**6**(3):174-184

[17] Koh EM, Kim HJ, Kim S, Choi WH, Choi YH, Ryu SY, et al. Modulation of macrophage functions by compounds isolated from *Zingiber officinale*. Planta Medica. 2009;**75**:148-151. DOI: 10.1055/s-0028-1088347

[18] Cho KJ, Kim JW, Choi IL, Kim JB, Hwang YS. Isolation, identification and determination of antioxidant in ginger (*Zingiber officinale*) rhizome. Agricultural Chemistry and Biotechnology. 2001;**44**:12-15. DOI: 10.1017/S1742170510000189

[19] Grzanna R, Lindmark L, Frondoza CG. Ginger—An herbal medicinal product with broad anti-inflammatory actions. Journal of Medicinal Food. 2005;**8**(2):125-132. DOI: 10.1089/jmf.2005.8.125.11

[20] Bernard MM, McConnery JR, Hoskin DW. [10]-Gingerol, a major phenolic constituent of ginger root, induces cell cycle arrest and apoptosis in triple-negative breast cancer cells. Experimental and Molecular Pathology. 2017;**102**:370-376. DOI: 10.1016/j.yexmp.2017.03.006

[21] Chen H, Lv L, Soroka D, Warin RF, Parks TA, Hu Y, et al. Metabolism of [6]-shogaol in mice and in cancer cells. Drug Metabolism and Disposition. 2012;**40**(4):742-753. DOI: 10.1124/dmd.111.043331

[22] Tahir AA, Sani NFA, Murad NA, Makpol S, Ngah WZW, Yusof YAM. Combined ginger extract & Gelam honey modulate Ras/ERK and PI3K/AKT pathway genes in colon cancer HT29 cells. Nutrition Journal. 2015;**14**:31. DOI: 10.1186/s12937-015-0015-2

[23] Kukic J, Petrovic S, Niketic M. Antioxidant activity of four endemic Stachys taxa. Biological and Pharmaceutical Bulletin. 2006;**29**:725-729. DOI: 10.1248/bpb.29.725

[24] Dai J, Mumper R. Plant phenolics: Extraction, analysis and their antioxidant and anticancer properties. Molecules. 2010;**15**:7313-7352. DOI: 10.3390/molecules15107313

[25] Takahashi M, Inouye S, Abe S. Anti-Candida and radical scavenging activities of essential oils and oleoresins of *Zingiber officinale* Roscoe and essential oils of other plants belonging to the family Zingiberaceae. Drug Discoveries and Therapeutics. 2011;**5**(5):238-245. DOI: 10.5582/ddt.2011.v5.5.238

[26] Sasidharan I, Menon AN. Comparative chemical composition and antimicrobial activity fresh & dry ginger oils (*Zingiber officinale* Roscoe). International Journal of Current Pharmaceutical Research. 2010;**2**(4):40-43

[27] Yuva B. Total antioxidant activity and antimicrobial potency of the essential oil and oleoresin of *Zingiber officinale* Roscoe. Asian Pacific Journal

of Tropical Diseases. 2014;**4**(1):40-44. DOI: 10.1016/S2222-1808(14)60311-X

[28] Zhang F, Zhang J, Qu J, Zhang Q, Prasad C, Wei Z. Assessment of anti-cancerous potential of 6-gingerol (Tongling white ginger) and its synergy with drugs on human cervical adenocarcinoma cells. Food Chemistry and Toxicology. 2017;**109**:910-922. DOI: 10.1016/j.fct.2017.02.038

[29] Cotelle N. Role of flavonoids in oxidative stress. Current Topics Medicinal Chemistry. 2001;**1**:569-590. DOI: 10.2174/1568026013394750

[30] Zhang G, Nitteranon V, Chan LY, Parkin KL. Glutathione conjugation attenuates biological activities of 6-dehydroshogaol from ginger. Food Chemistry. 2013;**140**:1-8

[31] Tomasz B, Anna O. Extraction methods for the isolation of isoflavonoids from plant material. Open Chemistry. 2017;**15**:34-45. DOI: 10.1515/chem-2017-0005

[32] Ghayur MN, Gilani AH. Pharmacological basis for the medicinal use of ginger in gastrointestinal disorders. Digestive Diseases and Sciences. 2005;**50**(10):1889-1897. DOI: 10.1007/s10620-005-2957-2

[33] Khaki AA, Khaki A. Antioxidant effect of ginger to prevent lead-induced liver tissue apoptosis in rat. Journal of Medicinal Plants Research. 2010;**4**: 1492-1495. DOI: 10.5897/JMPR09.397

[34] Suekawa M, Ishige A, Yuasa K, Sdo K, Aburada M, Hosoya E. Pharmacological studies on ginger. I. Pharmacological actions on pungent constituents, (6)-gingerol and (6)-shogaol. Journal of Pharmacobiodyn. 1984;**7**(11):836-848. DOI: 10.1248/bpb1978.7.836

[35] Kim JK, Kim Y, Na KM, Surh YJ, Kim TY. [6]-Gingerol prevents UVB-induced ROS production and COX-2 expression 'in vitro' and 'in vivo. Free Radical Research. 2007;**41**(5):603-614. DOI: 10.1080/10715760701209896

[36] Asnani V, Verma R. Antioxidative effect of rhizome of *Zingiber officinale* on paraben induced lipid peroxidation: An 'in vitro' study. Acta Poloniae Pharmaceutica. 2007;**64**(1):35-37

[37] Chohan M, Forster-Wilkins G, Opara E. Determination of the antioxidant capacity of culinary herbs subjugated to various cooking and storage processes using the ABTS (+) radical cation assay. Plant Foods for Human Nutrition. 2008;**63**(2):47-52. DOI: 10.1007/s11130-007-0068-2

[38] Tao Q, Xu Y, Lam R, et al. Diarylheptanoids and aminoterpenoid from the rhizome of *Zingiber officinale*: Antioxidant and cytoprotective properties. Journal of Natural Products. 2008;**71**(1):12-17. DOI: 10.1021/np070114p

[39] Herrmann K. Antioxidativ wiksame pflanzenphenole sowie carotinoide als wichtige inhalsstoffe von Gewu rzen. Gordian. 1994;**4**:113-117

[40] Chen IN, Ng CC, Wang CY, Chang TL. Lactic fermentation and antioxidant activity of Zingiberaceae plants in Taiwan. International Journal of Food Science and Nutrition. 2009;**60**(2):57-66. DOI: 10.1080/09637480802375531

[41] Portnoi G, Chng LA, Karimi-Tabesh L, Koren G, Tan MP, Einarson A. Prospective comparative study of the safety and effectiveness of ginger for the treatment of nausea and vomiting in pregnancy. American Journal of Obstetrics and Gynecology. 2003;**189**(5):1374-1377. DOI: 10.1067/s0002-9378(03)00649-5

[42] Heitmann K, Nordeng H, Holst L. Safety of ginger use in pregnancy:

Results from a large population-based cohort study. European Journal of Clinical Pharmacology. 2013;**69**(2):269-277. DOI: 10.1007/s00228-012-1331-5

[43] Stoilova I, Krastanov A, Stoyanova A, Denev P, Gargova S. Antioxidant activity of a ginger extract (*Zingiber officinale*). Food Chemistry. 2007;**102**(3):764-770. DOI: 10.1016/j.foodchem.2006.06.023

[44] Ippoushi K, Azuma K, Ito H, Horie H, Higashio H. [6]-Gingerol inhibits nitric oxide synthesis in activated J774.1 mouse macrophage and prevents peroxynitrite-induced oxidation and nitration reactions. Life Sciences. 2003;**73**(26):3427-3437. DOI: 10.1016/j.lfs.2003.06.022

[45] Zhang M, Viennois E, Prasad M, Zhang Y, Wang L, Zhang Z, et al. Edible ginger-derived nanoparticles: A novel therapeutic approach for the prevention and treatment of inflammatory bowel disease and colitis-associated cancer. Biomaterials. 2016;**101**:321-340. DOI: 10.1016/j.biomaterials.2016.06.018

[46] El-Ashmawy NE, Khedr NF, El-Bahrawy HA, Mansour HE. Ginger extract adjuvant to doxorubicin in mammary carcinoma: Study of some molecular mechanisms. European Journal of Nutrition. 2018;**57**(3):981-989. DOI: 10.1007/s00394-017-1382-6

[47] Liu CM, Kao CL, Tseng YT, Lo YC, Chen CY. Ginger phytochemicals inhibit cell growth and modulate drug resistance factors in docetaxel resistant prostate cancer cell. Molecules. 2017;**22**(9):1477. DOI: 10.3390/molecules22091477

[48] Liu Y, Whelan RJ, Pattnaik BR, Ludwig K, Subudhi E, Rowland H, et al. Terpenoids from *Zingiber officinale* (ginger) induce apoptosis in endometrial cancer cells through the activation of p53. PLoS One. 2012;**7**(12):153-178. DOI: 10.1371/journal.pone.0053178

[49] Brahmbhatt M, Gundala SR, Asif G, Shamsi SA, Aneja R. Ginger phytochemicals exhibit synergy to inhibit prostate cancer cell proliferation. Nutrition and Cancer. 2013;**65**(2):263-272. DOI: 0.1080/01635581.2013.749925

[50] Dias MC, Spinardi-Barbisan AL, Rodrigues MA, DeCamargo JL, Teran E, Barbisan LF. Lack of chemopreventive effects of ginger on colon carcinogenesis induced by 1,2-dimethylhydrazine in rats. Food Chemistry and Toxicology. 2006;**44**(6):877-884. DOI: 10.1016/j.fct.2005.11.015

[51] Habib SH, Makpol S, Abdul-Hamid NA, Das S, Ngah WZ, Yusok YA. Ginger extract (*Zingiber officinale*) has anti-cancer and anti-inflammatory effects on ethionine-induced hepatoma in rats. Clinics. 2008;**63**(6):807-813. DOI: 10.1590/s1807-59322008000600017

[52] Jiang Y, Turgeon DK, Wright BD, Sidahmed E, Ruffin MT, Brenner DE, et al. Effect of ginger root on cyclooxygenase-1 and 15-hydroxyprostaglandin dehydrogenase expression in colonic mucosa of humans at normal and increased risk for colorectal cancer. European Journal of Cancer Prevention. 2013;**22**(5):455-460. DOI: 10.1097/CEJ.0b013e32835c829b

[53] Kim EC, Min JK, Kim TY, Lee SJ, Yang HO, Han S, et al. [6]-Gingerol, a pungent ingredient of ginger inhibits angiogenesis 'in vitro' and 'in vivo'. Biochemical and Biophysical Research Communications. 2005;**335**(2):300-308. DOI: 10.1016/j.bbrc.2005.07.076

[54] Ling H, Yang H, Tan SH, Chui WK, Chew EH. 6-Shogaol, an active constituent of ginger, inhibits breast cancer cell invasion by reducing matrix metalloproteinase-9 expression via blockade of nuclear

factor-κB activation. British Journal of Pharmacology. 2010;**161**(8):1763-1777. DOI: 10.1111/j.1476-5381.2010. 00991.x

[55] Dugasani S, Pichika MR, Nadarajah VD, Balijepalli MK, Tandra S, Korlakunta JN. Comparative antioxidant and anti-inflammatory effects of [6]-gingerol, [8]-gingerol, [10]-gingerol and [6]-shogaol. Journal of Ethno-pharmacology. 2010;**127**(2):515-520. DOI: 10.1016/j.jep.2009.10.004

[56] Gull I, Saeed M, Shaukat H, Aslam SM, Samra ZQ, Athar AM. Inhibitory effect of *Allium sativum* and *Zingiber officinale* extracts on clinically important drug resistant pathogenic bacteria. Annals of Clinical Microbiology and Antimicrobials. 2012;**11**:8. DOI: 10.1186/1476-0711-11-8

[57] Srivastava KC. Aqueous extracts of onion, garlic and ginger inhibit platelet aggregation and alter arachidonic acid metabolism. Biomedica Biochimica Acta. 1984;**43**(8-9):335-346

[58] Yamahara J, Rong HQ, Naitoh Y, Kitani T, Fujimura H. Inhibition of cytotoxic drug-induced vomiting in suncus by a ginger constituent. Journal of Ethno-pharmacology. 1989;**27**(3):353-355. DOI: 10.1016/0378-8741(89)90010-x

[59] Malhotra S, Singh AP. Medicinal properties of ginger (*Zingiber officinale* Roscoe). Natural Products Radiance. 2003;**2**(6):296-300

[60] Nanjundaiah S, Annaiah H, Dharmesh S. Gastroprotective effect of ginger rhizome (*Zingiber officinale*) extract: Role of gallic acid and cinnamic acid in H^+, K^+ATPase/H. pylori inhibition and antioxidative mechanism. Evidence-Based Complementary and Alternative Medicine. 2011;**1**(1):1-13. DOI: 10.1093/ecam/nep060

[61] Chaiyakunapruk N, Kitikannakorn N, Nathisuwan S, Leeprakobboon K, Leelasettagool C. The efficacy of ginger for the prevention of postoperative nausea and vomiting: A meta-analysis. American Journal of Obstetrics and Gynaecology. 2006;**194**(1):95-99. DOI: 10.1016/j.ajog.2005.06.046

[62] Sontakke S, Thawani V, Naik MS. Ginger as an antiemetic in nausea and vomiting induced by chemotherapy; a randomized, cross-over, double blind study. Indian Journal of Pharmacology. 2003;**35**(1):32-36

[63] Rajesh KM, Anil K, Ashok K. Pharmacological activity of *Zingiber officinale*. International Journal of Pharmaceutical and Chemical Sciences. 2012;**1**(3):1073-1078

[64] Ajay M, Gilani AU, Mustafa MR. Effect of flavonoids on vascular smooth muscles of the isolated rat thoracic aorta. Life Sciences. 2003;**74**(5):603-612. DOI: 10.1016/j.lfs.2003.06.039

[65] Fischer-Rasmussen W, Kjaer SK, Dahl C, Asping U. Ginger treatment of hyperemesis gravida um. European Journal of Obstetrics, Gynecology and Reproductive Biology. 1991;**38**(1):19-24. DOI: 10.1016/0028-2243(91)90202-v

[66] Jung HW, Yoon CH, Park KM, Han HS, Park YK. Hexane fraction of *Zingiberis rhizoma* crudus extract inhibits the production of nitric oxide and proinflammatory cytokines in LPS-stimulated BV2 microglial cells via the NFkappaB pathway. Food Chemistry and Toxicology. 2009;**47**(6):1190-1197. DOI: 10.1016/j.fct.2009.02.012

[67] Srivastava KC, Mustafa T. Ginger (*Zingiber officinale*) in rheumatism and musculoskeletal disorders. Medical Hypothesis. 1992;**39**(4):342-348. DOI: 10.1016/0306-9877(92)90059-l

[68] Janet LF, Jennifer BF, Janice NO, Jianling C, Huaping Z, Barbara NT. Anti-inflammatory effects of the essential oils of ginger (*Zingiber*

officinale Roscoe) in experimental rheumatoid arthritis. PharmaNutrition. 2016;**4**(3):123-131. DOI: 10.1016/j.phanu.2016.02.004

[69] Flynn DL, Rafferty MF, Boctor AM. Inhibition of 5-hydroxyeicosatetraenoic acid (5-HETE) formation in intact human neutrophils by naturally occurring diarylheptanoids: Inhibitory activities of curcuminoids and yakuchinones. Prostaglandins Leukotrienes and Medicine. 1986;**22**(3):357-360

[70] Kiuchi F, Iwakami S, Shibuya M, Hanaoka F, Sankawa U. Inhibition of prostaglandin leukotriene biosynthesis by gingerols and diarylheptanoids. Chemical and Pharmaceutical Bulletin. 1992;**40**(2):387-391. DOI: 10.1248/cpb.40.387

[71] Chrubasik S, Pittler MH, Roufogalis BD. *Zingiberis rhizoma*: A comprehensive review on the ginger effect and efficacy profiles. Phytomedicine. 2005;**12**(9):684-701. DOI: 10.1016/j.phymed.2004.07.009

[72] Giriraju A, Yunus GY. Assessment of antimicrobial potential of 10% ginger extract against *Streptococcus mutans*, *Candida albicans*, and *Enterococcus faecalis*: An 'in vitro' study. Indian Journal of Dental Research. 2013;**24**(4):397-400. DOI: 10.4103/0970-9290.118356

[73] Sivasothy Y, Chong W, Hamid A, Eldeen IM, Sulaiman S, Awang K. Essential oil of *Zingiber officinale* var rubrum theilade and their antibacterial activities. Food Chemistry. 2011;**124**(2):514-517. DOI: 10.1016/j.foodchem.2010.06.062

[74] Norajit K, Laohakunjit N, Kerdchoenchuen O. Antibacterial effect of five Zingiberaceae essential oils. Molecules. 2007;**12**:2047-2060

[75] Ody P. The Complete Guide Medicinal Herbals. London: Dorling Kindersley; 2000. p. 75

[76] Chakotiya AS, Tanwar A, Narula A, Sharma RK. *Zingiber officinale*: Its antibacterial activity on *Pseudomonas aeruginosa* and mode of action evaluated by flow cytometry. Microbial Pathogenesis. 2017;**107**:254-260. DOI: 10.1016/j.micpath.2017.03.029

[77] Gupta S, Ravishankar S. A comparison of the antimicrobial activity of garlic, ginger, carrot and turmeric pastes against *Escherichia coli* O157:H7 in laboratory buffer and ground beef. Foodborne Pathogens and Disease. 2005;**2**:330-340. DOI: 10.1089/fpd.2005.2.330

[78] Wang W, Li CY, Wen XD, Li P, Qi LW. Simultaneous determination of 6-gingerol, 8-gingerol, 10-gingerol and 6-shogaol in rat plasma by liquid chromatography-mass spectrometry: Application to pharmacokinetics. Journal of Chromatography B: Analytical Technologies in the Biomedical and Life Sciences. 2009;**877**:671-679. DOI: 10.1016/j.jchromb.2009.01.021

[79] Sowley ENK, Kankam F, Afari D. Evaluation of neem (*Azadirachta indica*) seed and ginger (*Zingiber officinale*) as potential control agents of yam (*Dioscorea rotundata* Poir.) tuber rot fungi. Archives of Phytopathology and Plant Protection. 2013;**46**(17):2117-2124. DOI: 10.1080/03235408.2013.785659

[80] Yeni IJ. Inhibitory effects of two indigenous plant extracts (*Zingiber officinale* and *Ocimum gratissimum*) on postharvest yam (*Dioscorea rotundata* Poir.) rot 'in vitro'. Journal of American Science. 2011;**7**:43-47

[81] Ficker C, Smith ML, Akpagana K, Gbeassor M, Zhang J, Durst T, et al. Bioassay-guided isolation and identification of antifungal compounds from ginger. Phytotherapy Research. 2003;**17**(8):897-902. DOI: 10.1002/ptr.1335

[82] Imanishi N, Andoh T, Mantani N, et al. Macrophage mediated inhibitory effect of *Zingiber officinale* Roscoe, a traditional oriental herbal medicine, on the growth of influenza A/Aichi/2/68 virus. American Journal of Clinical Medicine. 2006;**34**:157-169

[83] Schnitzler P, Koch C, Reichling J. Susceptibility of drug-resistant clinical herpes simplex virus type 1 strains to essential oils of ginger, thyme, hyssop, and sandalwood. Anti-Microbial Agents Chemotherapy. 2007;**51**:1859-1862. DOI: 10.1128/AAC.00426-06

[84] Nerilo SB, Rocha GHO, Tomoike C, Mossini SAG, Grespan R, Mikcha JMG, et al. Antifungal properties and inhibitory effects upon aflatoxin production by *Zingiber officinale* essential oil in Aspergillus flavus. International Journal of Food Science and Technology. 2016;**51**:286-292

[85] Moon Y, Lee H, Lee S. Inhibitory effects of three monoterpenes from ginger essential oil on growth and aflatoxin production of *Aspergillus flavus* and their gene regulation in aflatoxin biosynthesis. Applied Biology and Chemistry. 2018;**61**:243-250. DOI: 10.1007/s13765-018-0352-x

[86] Supreetha S, Sharadadevi M, Sequeira P, Jithesh J, Shreyas T, Amit M. Antifungal activity of ginger extract on *Candida albicans*: An in-vitro study. Journal of Dental Sciences and Research. 2011;**2**(2):1-5

[87] Okigbo RN, Nmeka IA. Control of yam tuber rot with leaf extracts of *Xylopia aethiopica* and *Zingiber officinale*. African Journal of Biotechnology. 2005;**4**(8):804-807

[88] Rizvi SAH, Hussain S, Rehman SU, Jaffar S, Rehman MFU. Efficacy of ecofriendly botanical extracts of ginger (*Zingiber officinale*), garlic (*Allium sativum*) and tobacco (*Nicotiana tabacum* L) for the control of cabbage looper (*Trichoplusia binotalis*) under agro ecological conditions of Peshawar, Pakistan. Journal of Entomology and Zoology Studies. 2016;**4**(1):88-90

[89] Li F, Nitteranon V, Tang X, Liang J, Zhang G, Parkin KL, et al. *In vitro* antioxidant and anti-inflammatory activities of 1-dehydro-[6]-gingerdione, 6-shogaol, 6-dehydroshogaol and hexahydrocurcumin. Food Chemistry. 2012;**135**(2):332-337. DOI: 10.1016/j.foodchem.2012.04.145

[90] Kumar A, Goyal R, Kumar S, Jain S, Jain N, et al. Estrogenic and anti-Alzheimer's studies of *Zingiber officinalis* as well as *Amomum subulatum* Roxb.: The success story of dry techniques. Medicinal Chemistry Research. 2015;**24**(3):1089-1097

[91] Duarte MC. Antileishmanial activity and mechanism of action from a purified fraction of *Zingiber officinalis* Roscoe against *Leishmania amazonensis*. Experimental Parasitology. 2016;**166**:21-28. DOI: 10.1016/j.exppara.2016.03.026

[92] Isirima CB, Umesi N, Nnah MB. Comparative studies on effects of garlic (*Allium sativum*) and ginger (*Zingiber officinale*) extracts on cowpea insect pests attack. World Rural Observations. 2010;**2**(2):65-71. Available at: http://www.sciencepub.net/rural

[93] Sridhar S, Arumugasamy S, Saraswathy H, Vijayalakshmi K. Organic Vegetable Gardening. Chennai: Centre for Indian Knowledge System; 2002

[94] Amuji CF, Echezona BC, Dialoke SA. Extraction fractions of ginger (*Zingiber officinale* Roscoe) and residue in the control of field and storage pests. Journal of Agricultural Technology. 2012;**8**(6):2023-2031

[95] Sharma M, Shukla S. Hypoglycaemic effect of ginger. The Journal of Research in Indian Yoga and Homoeopathy. 1977;**12**:127-130

[96] Mascolo N, Jain R, Jain SC, Capasso F. Ethno-pharmacologic investigation of ginger (*Zingiber officinale*). Journal of Ethnopharmacology. 1989;**27**(1-2):129-140. DOI: 10.1016/0378-8741(89)90085-8

[97] Okamoto M, Irii H, Tahara Y, Ishii H, Hirao A, Udagawa H, et al. Synthesis of a new [6]-gingerol analogue and its protective effect with respect to the development of metabolic syndrome in mice fed a high-fat diet. Journal of Medical Chemistry. 2011;**54**(18):6295-6304. DOI: 10.1021/jm200662c

[98] Lin, Chen CY, Chung LY, Yen CM. Larvicidal activities of ginger (*Zingiber officinale*) against *Angiostrongylus cantonensis*. Acta Tropica. 2010;**115**(12):69-76. DOI: 10.1016/j.actatropica.2009.12.007

[99] Park G, Kim HG, Ju MS, Ha SK, Park Y, Kim SY, et al. 6-Shogaol, an active compound of ginger, protects dopaminergic neurons in Parkinson's disease models via anti-neuroinflammation. Acta Pharmacolologica Sinica. 2013;**34**:1131-1139

[100] Huh E, Lim S, Kim HG, Ha SK, Park H, Huh Y, et al. Ginger fermented with *Schizosaccharomyces pombe* alleviates memory impairment via protecting hippocampal neuronal cells in amyloid beta (1-42) plaque injected mice. Food Function. 2018;**9**:171-178. DOI: 10.1039/C7FO01149K

[101] Saraswat M. Antiglycating potential of *Zingiber officinalis* and delay of diabetic cataract in rats. Molecular Vision. 2010;**16**(165-66):1525-1537

[102] Nasri H, Nematbakhsh M, Ghobadi S, Ansari R, Shahinfard N, Rafieian-Kopaei M. Preventive and curative effects of ginger extract against histopathologic changes of gentamicin-induced tubular toxicity in rats. International Journal of Preventive Medicine. 2013;**4**(3):316-321

[103] Kubra IR, Murthy PS, Rao LJ. In vitro antifungal activity of dehydrozingerone and its fungitoxic properties. Journal of Food Sciences. 2013;**78**(1):64-69. DOI: 10.1111/j.1750-3841.2012.03009.x

Chapter 7

Ginger (*Zingiber officinale*) Antimicrobial Potential: A Review

Amanda Mara Teles, Bianca Araújo dos Santos, Cleidiane Gomes Ferreira, Adenilde Nascimento Mouchreck, Kátia da Silva Calabrese, Ana Lucia Abreu-Silva and Fernando Almeida-Souza

Abstract

Zingiber officinale Roscoe, commonly known as gengibre, ajengibre, jengibre dulce (Brazil, Argentina, and Spain), ginger (United States and England), and gingembre (France), is a perennial herbaceous plant that produces a fleshy and articulated rhizome, with rough brownish epidermis. As a medicinal plant, ginger is one of the oldest and most popular in the world. Several properties of the ginger have been verified in scientific experiments, with emphasis to the antimicrobial activity. Ginger essence oil has been investigated by several in vitro microbiological techniques, in which most of its essential oils presented antimicrobial activity against all selected bacteria. The antimicrobial effect is attributed mainly to several phytochemicals, such as camphene, phellandrene, zingiberene, and zingerone. This review provides an overview of the experimental evidence for the antimicrobial potential of *Z. officinale.*

Keywords: essential oil, chemical composition, ginger, gengibre

1. Introduction

Vegetable kingdom organisms are the major contributors to the significant number of organic substances in nature. Plants have enormous potential to biosynthesize the most varied types of molecular structures that perform various functions in your body. The substances responsible for ensuring the cells development and maintenance are called primary metabolites. From these compounds, through very complex biosynthetic routes, plants produce secondary metabolites, which help in the defense and adaptation of plants to the environment.

Composed of several secondary metabolites synthesized by plants, we highlight the essential oils that are characterized by being a complex mixture of low molecular weight liposoluble constituents with strong aroma. Essential oils stand out for their great therapeutic and economic importance, occupying a preponderant place in the pharmaceutical, cosmetic, and agri-food industries due to their high biological activity [1].

Although plants have been used since ancient times for spice and medicinal purposes, only in recent decades research has been intensifying for application of these compounds in food preservation and control of diseases of microbial origin.

Nowadays, there is a serious problem of bacterial resistance to commercially available antibiotics that occurs due to the wide distribution of antimicrobials and easy access to consumption by the population, which leads to indiscriminate use and self-medication. The uncertain diagnosis, the absence of a rational program for antimicrobial use, and subdoses of antimicrobial are also factors that contribute to the increased prevalence of drug-resistant microorganisms, rendering antibiotics ineffective [2].

Assuming the resistance of microorganisms to available drugs, the toxicity of synthetic antimicrobials, and the growing consumer awareness of the use of environmentally safe and health-friendly products, natural products emerge as a potential alternative for the replacement of synthetic antimicrobial agents.

One of the largest sources of research in this area is the evaluation of antimicrobial activity of plants popularly used for medicinal purposes. *Zingiber officinale* Roscoe, popularly known as ginger, is used in cooking, the pharmaceutical industry, and folk medicine to treat numerous conditions [3].

Thus, this review chapter aims to discuss the antimicrobial activity of ginger essential oil evaluated by various in vitro microbiological techniques against pathogenic microorganisms. This book chapter reviews the real contribution of ginger as a naturally occurring antimicrobial.

2. Methods

The bibliographic search was performed from May 2019 by a single researcher, searching for keywords such as antimicrobial, ginger, antibacterial, antifungal, *Zingiber officinale*, and their combinations, in PubMed and ScienceDirect. The productions were selected by reading and analyzing the titles and abstracts of all identified articles. After the initial screening, the selected studies were read, which allowed other texts that did not meet the review proposal to be excluded. The main information from the selected articles was synthesized in spreadsheets that guided the descriptive and critical analysis of the studies.

3. Results and discussion

3.1 Ginger plant

Ginger, scientifically named *Zingiber officinale* Roscoe, was first described in 1807 by the English botanist Willian Roscoe. It is a species in the Zingiberaceae family, from southwestern Asia and the Malay Archipelago, including over 1200 species and 53 genera [4].

Ginger has been known and used practically worldwide and in all medicines. It has been cultivated for thousands of years in China and India, reaching the West for at least 2000 years. The name of this genus, *Zingiber*, derives from a Sanskrit word meaning "horn-shaped" in reference to the protrusions on the surface of the rhizome. Ginger has several names, including gengibre, ajengibre, and jengibre dulce (Brazil, Argentina, and Spain), ginger (United States and England), and gingembre (France) [5, 6].

In Brazil, its cultivation was introduced shortly after the beginning of European colonization. However, only in the 1980s, with the introduction of giant rhizome varieties by Japanese farmers, ginger cultivation became effectively commercial in Brazil, especially in the coasts of Santa Catarina, Sao Paulo, and Paraná [7].

Ginger has a herbaceous habit, is perennial, produces articulated rhizome, and has adventitious roots and distal leaves, with the basilars reduced and floral bracts obliterated, each involving a single flower [8]. The ginger rhizome has an elongated, slightly flattened body, with a color ranging from yellow to bright brown leather, striated longitudinally, with endings known as "fingers" that arise obliquely from the rhizomes. Internally yellowish brown, it has a yellow endoderm, with numerous fibrovascular bundles and abundant oil cells. It presents pleasant and aromatic odor and strongly pungent taste [7].

As a medicinal plant, ginger is one of the oldest and most popular in the world. It is used to relieve symptoms of inflammation, rheumatic diseases, and gastrointestinal discomfort [9]. Its root has carminative, digestive, sweat, anti-influenza, and stimulating properties [8]. In gastronomy, ginger is used as a seasoning and flavoring, giving spicy and refreshing characteristics. It is a raw material for the manufacture of beverages and bakery products such as breads, cakes, cookies, and jams. In the cosmetics industry, its use is due to its fragrance [10].

Ginger has shown a variety of biological activities such as antifungal [11, 12], anti-inflammatory [13], antiviral [14], antimicrobial [3, 15], antioxidant [16], and antitumor [17–19]. Due to these properties, the use of rhizomes to obtain ginger essential oils, extracts, and concentrates has attracted interest from the pharmaceutical and food industries.

3.2 Chemical profile of ginger essential oils

Chemical analysis of ginger shows that it contains over 400 different compounds where the main components of ginger rhizomes are carbohydrates (50–70%), lipids (3–8%), terpenes, and phenolic compounds. Terpenes include zingiberene, β-bisabolene, α-farnesene, β-sesquifelenolene, and α-curcumene, while phenolic compounds include gingerol, paradols, and shogaol [20], as shown in **Table 1**.

In studies from 2006 to 2018 with *Z. officinale*, geranial and α-zingiberene were the major compounds in their chemical composition [12, 22, 27]. Significant quantities of the terpene family chemical constituents have also been reported [12, 22]. The least common constituent found in ginger essential oil is ar-curcumene [25, 27]. Compounds such as 1,8-cineole, eucalyptol, and 1,8-cinerol that are not very common in the chemical composition of *Zingiber officinale* have also been noted [21, 26, 28].

The variation in the composition of the essential oils obtained from this species may be due to genetic and/or environmental factors, plant age, and different extraction methods. The composition of essential oils directly influences their antimicrobial activity, as each secondary metabolite has a specific ability to break or penetrate the structure of the microorganism [24].

3.3 Antimicrobial assays with ginger

The most used methods for the determination of antimicrobial activity of *Z. officinale* vary among researches as can be seen in **Table 2**. The main methods are disk and well agar diffusion and agar and broth microdilution technique that determine the minimum inhibitory concentration (MIC) and minimum bactericidal concentration (MBC).

It was verified [16] that the *Z. officinale* essential oil tested by two methods showed strong inhibitory effects by well diffusion, demonstrating that the technique used can influence the result, while the agar diffusion test had less effect. As can be seen in **Table 2**, there is no standardization of ginger oil dilution, which can lead to uncertain results as it is an oil, and, because it is a less dense material than water, the oil cannot be diluted directly in the broth, which limits their miscibility

Reference	Major compounds	%
[12]	α-Zingiber	24.0
	Geraniale	15.0
	β-Phellandrene	8.0
[21]	Comphene	12.0
	β-Phellandrene	11.0
	1,8-cineal	10.0
	α-Zingiberene	7.0
[22]	α-Zingiberene	25.0
	β-Sesquiphellan	18.0
	β-Bisobeolene	12.5
[16]	Geraniale	26.0
	α-Zingiberene	9.5
	α-Farnesene	7.6
	Neral	7.4
[23]	Geraniale	16.0
	z-Citral	9.2
	β-Cedrene	8.6
	Geranyl acetate	8.4
[24]	Geraniale	26.0
	α-Zingibere	9.5
	Farnesene	7.6
	Neral	7.4
[25]	β-Sesquiphellandrene	27.0
	Caryophyllene	15.3
	Zingiberene	14.0
	α-Farnesense	10.5
[26]	ar-Curcumene	11.3
	Geraniale	11.0
	Camphene	5.0
	Eucalypto	3.0
[27]	α-Zingiberene	20.0
	ar-Curcumene	15.0
	β-Bisabalene	11.0
	β-Sesquiphellandrene	13.0
[28]	ar-Curamene	59.0
	1,8-Cinerol	8.0
	Citral	7.5
	α-Zingiberene	7.5

Table 1.
Chemical composition of different Zingiber officinale *essential oil described in literature.*

Method	Dilution	Reference
Disk and well diffusion agar	DMSO	[16]
Broth microdilution MIC and MBC	DMSO 5%	[21]
MIC—diffusion agar	Ethanol	[29]
Agar disk diffusion	Acetone	[30]
MIC—broth microdilution	Ginger essential oil	[31]
Agar disk diffusion	Essential oil	[32]
MIC—broth microdilution	Tween 80	[33]
Agar disk diffusion	Ginger essential oil	[34]
MIC-broth microdilution	Tween 126	[35]
Agar-agar diffusion	DMSO	[36]
Broth microdilution MIC and MBC	DMSO	[37]

DMSO, dimethyl sulfoxide; MIC, minimum inhibitory concentration; MBC, minimum bactericidal concentration.

Table 2.
Methods used to establish antimicrobial activity of Zingiber officinale *essential oil.*

in the test media. Therefore, a surfactant should be added, and we found that the most commonly used was DMSO [16, 21, 36, 37], tween [33, 35], some solvents like ethanol [29] and acetone [30], and even ginger oil [31, 32, 34].

The disk diffusion and well diffusion tests have been used to evidence antimicrobial activities, assuming that all components of the oil have the same solubility, but as verified in **Tables 3** and **4**, the diffusion of oil in the agar, during the test, may not diffuse into the agar, limiting the use of this method. However, the use of several methods to determine antimicrobial activity, as verified in [16], can directly interfere with the result. Although interference of chemical composition is possible, the MIC values found in several studies do not demonstrate a reproducibility using broth dilutions [21, 31, 33, 35, 37].

3.4 Antimicrobial activity of ginger

3.4.1 Antibacterial activity of ginger

Essential oils have a chemical composition rich in volatile and odorous secondary metabolites, mainly monoterpenes and sesquiterpenes. Several studies reported the antimicrobial properties of *Z. officinale* essential oil against various bacteria, as can be seen in **Table 3**.

A research showed that *Z. officinale* essential oil obtained by hydrodistillation verified that *L. monocytogenes* showed the highest sensitivity to oil when compared to other bacteria and presented the largest zone of inhibition (37 mm). Ginger essential oil has been shown to be active against the *V. alginolyticus* strain, despite the high MIC value range of 0.05–0.2 mg/mL reported [29].

The description of a moderate activity, with MIC values of 0.16–0.63 mg/mL, against Gram-positive bacteria indicates that Gram-negative bacteria are more resistant to *Z. officinale* essential oils compared to Gram-positive bacteria [21].

However, the essential oil showed activity against *Shigella* and *E. coli*, probably due to the presence of active constituents such as zingiberene, endoborneol, and gingerol [39]. The MIC value found for *K. pneumonia* (ATCC 13383) and *S. enterica* (ATCC 7251) strains was 1 mg/mL. These results are expected due to the constitution of the Gram-negative cell wall [16], although the effect of high sensitivity on Gram-negative strains such as *K. pneumonia* has been observed [39]. A survey of 15 strains of bacteria reported results that validate the use of *Z. officinale* as a medicine to treat diseases of possible infectious origin [9].

Reference	Country	Bacteria	MIC	MBC	Halo (mm)
[16]	India	*P. vulgaris*			18.4
		K. pneumoniae			20.5
[21]	Tunisia	*V. alginolyticus*		>25	
[23]	Brazil	*S. mutans*	250 μg/mL	500 μg/mL	
[29]	Saudi Arabia	*S. aureus*			15.8
		B. cereus			8.3
		E. coli			0.0
		S. typhi			0.0
		P. aeruginosa			11.2
[30]	Brazil	*S. enteritidis*			8.8
		L. plantarum			7.0
[32]	Saudi Arabia	*E. faecalis*	61.94%		
		P. aeruginosa	21.65%		
		E. coli	106.02%		
		Shigella	119.79%		
[37]	Canada	*S. pyogenes*	>1000 μg/mL	>1000 μg/mL	
[33]	Brazil	*L. monoctogenes*	4.7 μL/mL	9.4 μL/mL	
		S. aureus	2.3 μL/mL	4.7 μL/mL	
		E. coli O157:H7	9.4 μL/mL	18.7 μL/mL	
		S. typhimurium	9.4 μL/mL	18.7 μL/mL	
		P. aeruginosa	2.3 μL/mL	4.7 μL/mL	
[38]	India	*B. cereus*			9.11
		L. monocytogenes			9.00
		M. l nkluteus			6.86
		S. aureus			8.90
		E. coli			8.00
		S. typhimurium			6.61
[39]	Negeri Sembilan	*B. licheniformis*	0.16 mg/mL		
		B. spizizenii	0.24 mg/mL		
		E. coli	0.31 mg/mL		
		K. pneumoniae	0.47 mg/mL		
		P. stutzeri	0.63 mg/mL		
[40]	Mexico	*S. aureus*	0.25 mg/mL		
		S. epidemidis	0.5 mg/mL		
		E. faecalis	1.0 mg/mL		

Table 3.
Antibacterial activity of Zingiber officinale *essential oil.*

Reference	Fungi	MIC	Disk diffusion	
			Halo	Concentration
[16]	*A.flavus*		20.6 mm	6 µg/mL
	A. solani		66.3 mm	
	A. oryzae		51.3 mm	
	A. Níger		66.7 mm	
	F. moniliforme		100 mm	
[35]	*C. albicans*		25 mm	100 µg/mL
	G. candidum		21 mm	
	F. oxysporum		22 mm	
	A. flavus		20 mm	
[36]	*F. verticillioides*	2500 µg/mL		
[41]	*A. terrus*		50%	10 µL
	A. Niger		31.3%	
	A. flavus		87.5%	
	F. oxysporum		87.5%	
	C. palliscens		87.5%	
	T. roseum		100%	
	F. graminearum		62.5%	
	F. monoliforme		75%	
[42]	*Penicillium* spp	869.2 mg/mL		

Table 4.
Zingiber officinale *antifungal activity.*

A research conducted in Brazil with a substance (zerumbone) isolated from ginger essential oil showed its efficacy against *S. mutans*, resulting in 250 µg/mL MIC and 500 µg/mL MBC. Another investigation of the effect of oil against growth activity and biofilm formation of *S. pyogenes* showed MIC and MBC of 1 mg/mL [37].

We found that the studies reported in this review show that the antibacterial effect of essential oil has significant differences according to the collection site, its genetic and environmental composition of the plant, and extraction methods, as well as significant differences in the inhibition of Gram-positive and Gram-negative bacteria. Gram-positive strains are more sensitive, suggesting that the cell wall composed of a thick layer of peptidoglycan surrounding the cytoplasmic membrane would be the microbial target of essential oil [43].

However, the possibility of another target is not ruled out, as we found that, depending on the location, the oil tested demonstrates a better effect on Gram-negative, suggesting other microbial targets, such as the plasma membrane, since the constituents of essential oils have lipophilic properties that interact with membranes by changing their fluidity and permeability [44].

3.4.2 Antifungal activity of ginger

In the evaluation of antifungal activity, we found that antifungal tests with *Z. officinale* oil showed inhibitory effects against all fungal tested. Ginger oil was found to completely inhibit *F. moniliforme* growth at the highest concentration tested, and *Aspergillus* inhibition was also reported.

A study with oils obtained by different drying methods against six fungi (*Candida albicans*, *Geotrichum candidum*, *Trichophyton rubrum*, *Aspergillus flavus*, *Fusarium oxysporum*, and *Scopulariopsis brevicaulis*) revealed that hot-drying ginger exhibited potent antifungal activity except against *T. rubrum* and *S. brevicaulis* when the oil was obtained by drying indoors. In open-air drying, the oil showed antifungal activity only against *C. albicans* [35].

The activity against *Fusarium verticillioides* determined by broth dilution exhibited MIC of 2500 μg/mL, suggesting that ginger oil is capable of controlling *F. verticillioides* growth and subsequent fumonisin production [36]. Both essential oil and ginger resin totally inhibited (100%) *Fusarium moniliforme* [41]. Activity against other fungi showed moderated to good effect (**Table 4**).

The antimicrobial activity of ginger oil can be attributed to its constituent monoterpenes and sesquiterpenes, as they are capable of altering the permeability and fluidity of the plasma membrane of microorganisms. The lipophilic character of its hydrocarbon skeleton and the hydrophilic character of some of its functional groups confer this property [40].

Z. officinale essential oil contains considerable amounts of phenolic compounds (eugenol, shogaols, zingerone, gingerdiols, gingerols, etc.), which may be responsible for the observed effects, and has different chemotypes in which the efficiency can be attributed to the major compounds, although the possibility of a synergistic action of all constituents is not ruled out either [41].

4. Conclusions

The studies reported in this literature review made the determination of the species, the indication of the place of collection, and the extraction method, since these data are fundamental for adequate comparison of the results, as well as a secondary metabolite identification technique where we found that the most used techniques were gas chromatography (GC) and liquid chromatography (HPLC) to indicate the present compounds. Geographical location, oil extraction method, techniques, media types, dilution used in antimicrobial activity at different concentrations, and microorganisms can certainly lead to different results. Ginger essential oil has compounds that are present in varying proportions as verified in this review; therefore, there is no parameter for their composition as they have several chemotypes. The lack of oil standardization makes it difficult to compare the work done and to obtain an adequate result of the antimicrobial activity of the oil. However, numerous reports of antimicrobial activity, even with the various variables described above, lead us to believe that ginger essential oil has a potential antimicrobial activity to be explored, and further studies are needed to ensure this activity.

Author details

Amanda Mara Teles[1], Bianca Araújo dos Santos[1], Cleidiane Gomes Ferreira[1], Adenilde Nascimento Mouchreck[1], Kátia da Silva Calabrese[2], Ana Lucia Abreu-Silva[3] and Fernando Almeida-Souza[2]*

1 Federal University of Maranhão, São Luís, Brazil

2 Oswaldo Cruz Institute, Rio de Janeiro, Brazil

3 State University of Maranhão, São Luís, Brazil

*Address all correspondence to: fernandoalsouza@gmail.com

References

[1] Andrade MA. Óleos essenciais de Cinnamomum zeylanicum, *Cymbopogon nardus* e *Zingiber officinale*: Caracterização química, atividade antioxidante e antibacteriana [Dissertação (Mestrado em Agroquímica)]. Minas Gerais: Universidade Federal de Lavras; 2010. 101 p

[2] Mota LM, Vilar FC, Dias LBA, Nunes TF, Moriguti JC. Uso racional de antimicrobianos. Medicina (Ribeirão Preto). 2010;**43**(2):164-172

[3] Diemer AW. Ação antimicrobiana de Rosmarinusofficinalis e *Zingiber officinale* frente a *Escherichia coli* e *Staphylococcus aureus* em carne mecanicamente separada de frango [Dissertação (Mestrado em Biotecnologia)]. Lajeado, Rio Grande do Sul: Centro Universitário Univates; 2016. 67 p

[4] Lorenzi H, Matos FJA. Plantas medicinais no Brasil: Nativas e exóticas. Plantarum: Nova Odessa; 2002. 512 p

[5] Corrêa Junior C, Ming LC, Scheffer MC. Cultivo de plantas medicinais, condimentares e aromáticas. 2ª. ed. Jaboticabal: FUNEP; 1994. 151 p

[6] Morgan R. Enciclopédia das Ervas e Plantas Medicinais. Hemus: São Paulo; 1994

[7] Martins AGLA. Atividade antibacteriana dos óleos essenciais do manjericão (*Ocimum basilicum* Linnaeus) e do gengibre (*Zingiber officinale* Roscoe) frente a linhagens de Escherichia coli enteropatogênicas isoladas de hortaliças [Tese (Doutorada em Ciências dos Alimentos)]. João Pessoa: Universidade Federal do Paraíba; 2010. 110 p

[8] Soares RP. Atividade biológica dos óleos essenciais de gengibre, açafrão e louro sobre o fungo Aspergillus carbonarius [Dissertção (Mestrado em Agroquímica)]. Lavras, Minas Gerais: Universidade Federal de Lavras; 2009. 79 p

[9] Pfeiffer E, Heuschmid FF, Kranz S, Metzler M. Microsomal hydroxylation and glucuronidation of [6]-gingerol. Journal of Agricultural and Food Chemistry, Easton. 2006;**54**(23):8769-8774

[10] Mukkavilli R, Yang C, Singh Tanwar R, Ghareeb A, Luthra L, Aneja R. Absorption, metabolic stability, and pharmacokinetics of ginger phytochemicals. Molecules. 2017;**22**:553

[11] Freire JCP, Júnior JK d O, Silva D d F, de Sousa JP, Guerra FQS, de Oliveira Lima E. Antifungal activity of essential oils against *Candida albicans* strains isolated from users of dental prostheses. Evidence-based Complementary and Alternative Medicine. 2017;**2017**:1-9

[12] Ferreira FMD, Hirooka EY, Ferreira FD, Silva MV, Mossini SAG, Machinski M Jr. Effect of *Zingiber officinale* Roscoe essential oil in fungus control and deoxynivalenol production of *Fusarium graminearum* Schwabe in vitro. Food Additives & Contaminants: Part A. 2018;**35**(11):2168-2174

[13] Camargo LCS. Efeito antiin-flamatorio do extrato de Zingiber officinale aplicado por fonoforese sobre o edema de pata de ratos [Dissertação (Mestrado em Ciências Biológicas)]. São José dos Campos: Instituto de Pesquisa e Desenvolvimento, Universidade do Vale do Paraíba; 2006. 89 p

[14] Camero M, Lanave G, Catella C, Capozza P, Gentile A, Fracchiolla G, et al. Virucidal activity of ginger essentialoilagainstcaprinealphaherpes virus-1. Veterinary Microbiology. 2019;**230**:150-155

[15] Cutrim ESM, Teles AM, Mouchrek AN, Mouchrek Filho VE, Everton GO. Avaliação da Atividade Antimicrobiana e Antioxidante dos Óleos Essenciais e Extratos Hidroalcoólicos de Zingiber officinale (Gengibre) e *Rosmarinus officinalis* (Alecrim). Revista Virtual de Química. 2019;**11**(1):60-81

[16] Singh G, Kapoor IPS, Singh P, Heluani CS, Lampasona MP, Catalan CAN. Chemistry, antioxidant and antimicrobial investigations on essential oil and oleoresins of *Zingiber officinale*. Food and Chemical Toxicology, Oxford. 2008;**46**(10):3295-3302

[17] Dorai T, Aggarwal BB. Role of chemopreventive agents in cancer therapy. Cancer Letters. 2004;**215**:129-140

[18] Manju V, Nalini N. Chemopreventive efficacy of ginger, a naturally occurring anticarcinogen during the initiation, post-initiation stages of 1,2dimethylhydrazine-inducedcoloncancer. Clinica Chimica Acta. 2005;**358**:60-67

[19] Shukla Y, Singh M. Cancer preventive properties of ginger: A brief review. Food and Chemical Toxicology. 2007;**45**:683-690

[20] Grzanna R, Lindmark L, Frondoza CG. Ginger—An herbal medicinal product with broad anti-inflammatory actions. Journal of Medicinal Food. 2005;**8**(2):125-132

[21] Snuossi M, Trabelsi N, Taleb S, Dehmeni A, Flamini G, Feo V. *Laurus nobilis*, *Zingiber officinale* and *Anethum graveolens* essential oils: Composition, antioxidant and antibacterial activities against bacteria isolated from fish and shellfish. Molecules. 2016;**21**:1414. DOI: 10.3390/molecules21101414

[22] Varoni EM, Lo Faro AF, Sharifi-Rad J, Iriti M. Anticancer molecular mechanisms of resveratrol. Frontiers in Nutrition. 2016;**3**:8. DOI: 10.3389/fnut.2016.00008

[23] Silva M, Pinheiro C, Orlandi P, Pinheiro C, Pontesa J. Zerumbone from *Zingiber zerumbet* (L.) *smith*: A potential prophylactic and therapeutic agent against the cariogenic bacterium *Streptococcus mutans*. BMC Complementary and Alternative Medicine. 2018;**18**:301. DOI: 10.1186/s12906-018-2360-3010

[24] Chmit M, Kanaan H, Habib J, Abbass M, Mcheik A, Chokr A. Antibacterial and antibiofilm activities of polysaccharides, essential oil, and fatty oil extracted from *Laurus nobilis* growing in Lebanon. Asian Pacific Journal of Tropical Medicine. 2014:546-552. DOI: 10.1016/S1995-7645(14)60288-1

[25] Borah A, Sethi L, Sarkar S, Hazarika K. Effect of drying on texture and color characteristics of ginger and turmeric in a solar biomass integrated dryer. Journal of Food Process Engineering. 2017;**40**:e12264. DOI: 10.1111/jfpe.12310

[26] Mesomo MC, Corazza ML, Ndiaye PM, Dalla Santa OR, Cardozo L, de Paula Scheer A. Supercritical CO_2 extracts and essential oil of ginger (*Zingiber officinale* R.): Chemical composition and antibacterial activity. Journal of Supercritical Fluids. 2013;**80**:44-49. DOI: 10.1016/j.supflu.2013.03.031

[27] Wang Z, Wang L, Li T, Zhou X, Ding L, Yu Y, et al. Rapid analysis of the essential oils from dried Illicium verum Hook. f. and Zingiber officinale Rosc. by improved solvent-free microwave extraction with three types of microwave-absorption medium. Analytical and Bioanalytical Chemistry. 2006;**386**(6):1863-1868. DOI: 10.1007/s00216-006-0778-6

[28] Nogueira de Melo GA, Grespan R, Fonseca JP, Farinha TO, da Silva EL, Romero AL, et al. Inhibitory effects of ginger (*Zingiber officinale* roscoe) essential oil on leukocyte migration in vivo and in vitro. Journal of Natural Medicines. 2011;**65**:241-246. DOI: 10.1007/s11418-010-0479-5

[29] Mostafa A, Abdulaziz A, Al-Askar A, Khalid S, Turki M, Essam N, et al. Antimicrobial activity of some plant extracts against bacterial strains causing food poisoning diseases. Saudi Journal of Biological Sciences. 2018;**25**(2):361-366. DOI: 10.1016/j.sjbs.2017.02.004

[30] Ambrosio MS, Severino M, Alencar LM, Sousa M, Gloria M. Antimicrobial activity of several essential oils on pathogenic and beneficial bacteria. Industrial Crops and Products. 2017;**97**(2017):128-136. DOI: 10.1016/j.indcrop.0926-6690

[31] Chakotiya A, Tanwar A, Narula A, Sharma R. *Zingiber officinale*: Its antibacterial activity on *Pseudomonas aeruginosa* and mode of action evaluated by flow cytometry. Microbial Pathogenesis. 2017;**107**:254-260. DOI: 10.1016/j.micpath.2017.03.02-9 0882-4010

[32] Ashraf S, Al-Shammari E, Hussain T, Tajuddin S, Panda B. In-vitro antimicrobial activity and identification of bioactive components using GC–MS of commercially available essential oils in Saudi Arabia. Association of Food Scientists & Technologists. 2017;**54**(12):3948-3958. DOI: 10.1007/s13197-017-2859-2

[33] Tavares F, Cunha K, Fonseca L, Antunes M, Mello S, Fiorentini A, et al. Action of ginger essential oil (*Zingiber officinale*) encapsulated in proteins ultrafine fibers on the antimicrobial control in situ. International Journal of Biological Macromolecules. 2018;**118**(2018):107-115. DOI: 10.1016/j.ijbiomac.2018.06.079

[34] El-Shouny WA, Ali SS, Sun J, Samy SM, Ali A. Drug resistance profile and molecular characterization of extended spectrum beta-lactamase (ESβL)-producing *Pseudomonas aeruginosa* isolated from burn wound infections. Essential oils and their potential for utilization. Microbial Pathogenesis. 2018;**116**:301-312. DOI: 10.1016/j.micpath.2018.02.005

[35] Ghasemzadeh A, Jaafar H, Baghdadi A, Tayebi-Meigooni A. Formation of 6-, 8- and 10-Shogaol in ginger through application of different drying methods: Altered antioxidant and antimicrobial activity. Molecules. 2018;**23**:1646. DOI: 10.3390/molecules23071646

[36] Yamamoto-Ribeiro MMG, Grespan R, Kohiyama CY, Ferreira FD, Mossini SAG, Silva EL, et al. Effect of *Zingiber officinale* essential oil on *Fusarium verticillioides* and fumonisin production. Food Chemistry. 2013;**141**(3):3147-3152. DOI: 10.1016/j.foodchem.2013.05.144

[37] Wijesundara NM, Rupasinghe HPV. Essential oils from *Origanum vulgare* and *Salvia officinalis* exhibit antibacterial and anti-biofilm activities against streptococcus pyogenes. Microbial Pathogenesis. 2018;**13**(6):612-632. DOI: 10.1016/j.micpath.2018.02.026

[38] Bag A, Chattopadhyay RR. Evaluation of synergistic antibacterial and antioxidant efficacy of essential oils of spices and herbs in combination. PLoS One. 2015;**10**(7):e0131321. https://doi.org/10.1371/journal.pone.0131321

[39] Sivasothy Y, Chong WK, Hamid A, Eldeen IM, Sulaiman SF, Awang K. Essential oils of Zingiber officinale var. rubrum Theilade and their antibacterial activities. Food

Chemistry. 2011;**124**:514-517. DOI: 10.1016/j.foodchem.2010.06.062

[40] López EIC, Balcázar MFH, Mendoza JMR, Ortiz ADR, Melo MTO, Parrales RS, et al. Antimicrobial activity of essential oil of Zingiber officinale roscoe (Zingiberaceae). American Journal of Plant Sciences. 2017;**8**(07):1511

[41] Singh G, Maurya S, Catalan C, De Lampasona MP. Studies on essential oils, part 42: Chemical, antifungal, antioxidant and sprout suppressant studies on ginger essential oil and its oleoresin. Flavour and Fragrance Journal. 2005;**20**(1):1-6

[42] BELLIK Y. Total antioxidant activity and antimicrobial potency of the essential oil and oleoresin of *Zingiber officinale* roscoe. Asian Pacific Journal of Tropical Disease. 2014;**4**(1):40-44

[43] Burt S. Essential oils: Their antibacterial properties and potential applications in foods—A review. International Journal of Food Microbiology. 2004;**94**:223-253. DOI: doi.org/10.1016/j.ijfoodmicro.2004.03.022

[44] Berger RG. Bioactivity of essential oils and their components. In: Flavors and Fragrances: Chemistry, Bioprocessing, and Sustainability. Berlin: Springer; 2007. pp. 88-90. DOI: 10.1007/978-3-540-49339-6

Section 4

Pharmacological Potentials

Chapter 8

A Review of the Antidiabetic Activities of Ginger

Gloria Aderonke Otunola and Anthony Jide Afolayan

Abstract

Diabetes mellitus, a chronic metabolic disorder with major health care burden worldwide, is increasing, with 173 million adults being diabetic and over 8 million deaths recorded annually. Undesirable pathological conditions and high rates of secondary failure limit the use of current antidiabetic agents, thus, the need for more effective antidiabetic agents. Medicinal plants such as spices, rich in bioactive components that promote prevention and treatment of chronic conditions such as heart disease, cancer and Type-2 diabetes, are inexpensive with no side effects. The Zingiberaceae family, of which ginger is a member, consists of many species frequently cited for their antidiabetic and hypoglycemic properties. All important scientific literatures from 2000 to 2018 on the antidiabetic potentials of *Zingiber officinale* were evaluated. According to these studies, ginger exerts its antidiabetic effects through restorative effects on pancreatic β-cells, increasing insulin sensitivity, action and peripheral utilization of glucose. Other mechanisms include increased synthesis of hepatic glycogen through the enhancement of glycogen regulatory enzyme expression in the liver, inhibition of carbohydrate metabolizing enzymes, stimulation of pancreatic insulin release and inhibition of hepatic glucose production. Further studies, especially in humans are needed, more so, since ginger is one of the spices generally regarded as safe.

Keywords: spices, diabetes, ginger, pharmacology, mechanism of action

1. Introduction

Diabetes mellitus (DM) is the most common endocrine disorder that affects more than 100 million people worldwide. It is a heterogeneous group of diseases, all of which ultimately lead to an elevation of glucose in the blood (hyperglycemia) and loss of glucose in the urine as hyperglycemia increases. It is characterized by increased urine production (polyurea) excessive thirst (polydipsia) and excessive eating (polyphagia).

Diabetes mellitus is a chronic metabolic disorder of the endocrine system that is characterized by defects in impaired metabolism of glucose, lipid and protein as well as insulin secretion or insufficiency. Diabetes continues to be a major health care problem worldwide and its prevalence is expected to rise from the current 382–471 million individuals by 2035 [1, 2]. There are three main types of diabetes- Type 1 diabetes (T1D), which is an autoimmune disorder leading to the destruction of pancreatic beta-cells; Type 2 diabetes (T2D), which is much more common and primarily caused by impaired glucose regulation due to a combination of dysfunctional pancreatic beta cells and insulin resistance and gestational diabetes mellitus (GDM).

Different treatments, such as insulin, pharmacotherapy and diet therapies, which exert antidiabetic effects through different mechanisms, are currently used for the management of diabetes. Such mechanisms include stimulation of insulin secretion by sulfonylurea and meglitinides drugs, increase of peripheral absorption of glucose by biguanides and thiazolidinediones, delay in the absorption of carbohydrates from the intestine by alpha-glucosidase and reduction of hepatic gluconeogenesis by biguanides [3–5].

In spite of the appreciable progress that has been made in the management of diabetes through the use of conventional drugs and management strategies, diabetes and its complications continue to be a major medical problem and rising burden of disease. Most synthetic oral hypoglycemic agents available for the treatment of the disease have some disadvantages, including drug resistance, serious side effects, cannot be used during pregnancy, are toxic and also costly [6, 7].

Spices and herbs have played important roles in civilization and history of many nations of the world. Their flavor and pungency makes them indispensable in the preparation of palatable dishes; but beyond adding flavor, spices are reputed to possess several medicinal and pharmacological properties and hence find use in the preparation of a number of medicines. Spices can be the dried leaf (e.g., bay leaf), buds (cloves), bark (cinnamon), rhizome/root (ginger), berries (grains of pepper), seeds (cumin), or even the stigma of the flower (saffron) [8].

The Zingiberaceae plant family consists of many species used as culinary herbs and spices, frequently cited for their antidiabetic and hypoglycemic properties. Ginger (*Zingiber officinale*) belongs to this family, and has a long and wide history of usage both as a culinary spice and in traditional/alternative medicine. This study attempts to update the available scientific information on the antidiabetic and hypoglycemic potentials of ginger.

2. Methodology

Online published articles from Google Scholar, ScienceDirect, Scopus, ResearchGate, PubMed and SciELO were explored for data collection. For literature search, key words such as spices, diabetes, Zingiberaceae, ginger, in vivo, in vitro, pharmacological, medicinal, hypoglycemic and antidiabetic were used. The study reviewed all important literature from 2000 to 2018.

3. Results

The following sections describe various studies reporting the hypoglycemic and antidiabetic properties of ginger, phytochemical constituents responsible for these properties and its mechanisms of action.

3.1 Ginger (*Zingiber officinale* Roscoe)

The ginger (Zingiberaceae) family consists of 53 genera and over 1200 medicinal plants, typically tropical annuals or perennials, often with large rhizomes. This plant family is well-known for its medicinal values and is distributed widely throughout the tropics, particularly in Southeast Asia.

Ginger (**Figure 1**) has been used for thousands of years for the treatment of numerous ailments, such as colds, nausea, arthritis, migraines and hypertension. Several authors have reviewed the medicinal, chemical, and pharmacological properties of ginger [9–13]. Ginger is recognized by the U.S. Food and Drug

Figure 1.
Zingiber officinale (ginger) rhizomes. Source: www.amazon.com/Culinary-Ginger-Zingiber-Officinalecooking/dp/B01NAJQ6C6 [28].

Administration (FDA) as a food additive that is "generally recognized as safe" [14] and has a long history of use as a culinary spice and in herbal medicine to treat many ailments.

Ginger helps to relieve various inflammatory disorders like gout, osteoarthritis and rheumatoid arthritis; as an analgesic, hypoglycemic, cardiotonic, antiemetic, antimicrobial and antifungal. Other uses include as an antidiabetic, antidyslipidemia, hypotensive, vasodilator, antiobesity and anticancer agent [15–18].

In addition, the spice has shown prominent protective effects on diabetic liver, kidney, eye, and neural system complications [19]. Several experimental studies have reported that administration of ginger extracts significantly decreased blood glucose level in Type 1- and Type 2-induced diabetic animals in a dose-dependent pattern [20–27].

3.2 Antidiabetic properties of ginger

Lamuchi-Deli et al. [29] evaluated the effects of the hydroalcoholic extract of *Zingiber officinale* on arginase I activity and expression in the retina of streptozotocin-induced diabetic rats. The study showed that blood glucose concentration was significantly decreased, arginase I activity and expression was significantly ($P < 0.05$) down regulated and 400 mg/kg of the extract reduced significant elevation in body weight in diabetic rats compared to untreated diabetic controls ($P < 0.01$). Serum insulin was also significantly increased in diabetic rats treated with 400 mg/kg of the extract compared to diabetic controls ($P < 0.05$). The study suggested that ginger could be a promising therapeutic option for treating diabetes-induced vascular disorders.

A recent study by de Las Heras et al. [30] which evaluated the hypolipidemic- and insulin sensitizing effects of hydroethanolic extract of ginger in the liver of high-fat diet (HFD) fed rats, revealed that ginger extract improved lipid profile and attenuated the increase of plasma levels of glucose, insulin, and leptin in HFD rats. Wei et al. [31], also reported that 6-Paradol and 6-Shogaol, the pungent compounds of ginger, promoted glucose utilization in adipocytes and myotubes, and 6-paradol reduced blood glucose in HFD-fed mice. Al-Qudah et al. [32] reported that aqueous extract of ginger was effective in lowering serum glucose, restoration of hematological indices to normal and repair damaged pancreas in alloxan-induced diabetic rats.

In another study, Oludoyin and Adegoke [33] investigated the effect of ginger extracts on blood glucose in normal and streptozotocin-induced diabetic rats. The authors reported that the fasting blood glucose in diabetic rats was reduced to normal by both raw and cooked ginger extracts in a manner comparable to glibenclamide. Evaluation of the nutritional and antidiabetic activity of ginger powder, its aqueous and methanolic extract, as well as the essential oil in streptozotocin-induced diabetic rats [34] revealed reduction in levels of alanine and aspartate aminotransferase (ALT and AST), alkaline phosphatase (ALP), liver total lipid and cholesterol of diabetic rats; and increased levels of liver glycogen and triglyceride compared to positive control group. In the study, ginger oil showed the best antidiabetic activity, followed by ginger extracts. Again, another study reported that ginger extract administered at 200 mg/kg/day/kg body weight for 10 weeks to male Sprague-Dawley diabetic rats, exhibited protective activity against insulin resistance [34].

Al-Noory et al. [35] showed that fresh ginger extracts led to decrease in the levels of total cholesterol (TC) and low density lipoprotein (LDL) in the serum of alloxan-induced diabetic rats, compared with the control groups; and previous extracts caused reduction in LDL to levels comparable to normal group and equal to the effect of atorvastatin given at a dosage of 10 mg/day. Similarly, oral administration of aqueous ginger extract to streptozotocin (STZ)-induced diabetic rats for a period of 30 days was reported to give a dose-dependent antihyperglycemic effect, 68% decrease in plasma glucose level at a daily dose of 500 mg/kg body weight, indicating that ginger is a potential phytomedicine for the treatment of diabetes [36]. Iranloye et al. [37] also showed that ginger effectively reduced fasting blood glucose, malondialdehyde levels and enhanced insulin sensitivity in alloxan-induced and insulin-resistant diabetic rats compared to control rats.

Treatment of streptozotocin-induced Type I diabetic rats with *Z. officinale* juice (4 mL kg^{-1}, p.o. daily for 6 weeks) was reported to produce a significant increase in insulin levels, decrease in fasting glucose levels, as well as significant decrease in the area under the curve of glucose in an oral glucose tolerance test [38]. According to Nammi et al. [23], treatment with an ethanolic extract of ginger at doses of 100, 200, and 400 mg/kg for 6 weeks, significantly reduced the marked increase in body weight, serum glucose, insulin, total cholesterol, LDL cholesterol, triglycerides, free fatty acid and phospholipids induced by high-fat diet.

The study conducted by Al-Amin et al. [39] on the antidiabetic and hypolipidemic properties of ginger (*Zingiber officinale*) in streptozotocin-induced diabetic rats revealed that raw ginger at a dose of 500 mg/kg, was significantly effective in lowering serum glucose, cholesterol and triacylglycerol as well as reduction in urine protein (reversal of diabetic proteinuria) levels, of diabetic rats. Ethanolic extracts of *Zingiber officinale* (200 mg/kg) given orally for 20 days was reported to produce significant antihyperglycemic effect ($P < 0.01$) in diabetic rats, while also lowering serum total cholesterol and triglycerides, coupled with increased HDL-cholesterol levels when compared with pathogenic diabetic rats [40].

Hypoglycemic effect of ginger (4 and 8 g/kg), administered intraperitoneally to rats after 30 min of diabetes induction, with the effect being more pronounced after 2 h has been reported. In another study, Otunola and Afolayan [41], showed that aqueous extract of a spice mixture containing ginger at 500 mg/kg body weight extract significantly ($p < 0.05$) lowered the elevated fasting blood glucose, lipid and hematological indices of alloxan-induced diabetic rats at equipotent level with glibenclamide.

Several in vitro hypoglycemic potentials of ginger and its bioactive constituents especially 6-shogaol have been reported (**Table 1**).

The capacity of ginger for hypoglycemic, antidiabetic, insulogenic, better glucose tolerance, increased serum insulin levels, reduction in elevated lipid levels, and prevention of weight loss associated with diabetes in human diabetic patients have also been reported (**Table 2**).

3.3 Phytochemical components of *Zingiber officinale*

GC-MS profiling of diethyl extracts as reported by Koch et al. [46] showed the presence of monoterpenes such as (α-pinene, camphene, myrcene, and α-phellandrene), oxygenated monoterpenes (geranial, citronellal, neral, linalool, borneol, and α-terpineol), and sesquiterpenes (α-and β-farnesene, ar-curcumene, zingiberene, zingiberenol, copaene, or cadinene). The most abundant substances in the extracts were α-zingiberene (37.9%), β-sesquiphellandrene (11.4%), (E,E)-α-farnesene (9.6%), geranial (8.2%), ar-curcumene (6.3%), and γ-terpinene (5.1%).

Similarly, Sharma et al. [47] reported that the essential oil of fresh ginger rhizome was characterized by high percentage of sesquiterpenes (66.66%), monoterpenes (17.28%) and aliphatic compounds (13.58%). The predominant sesquiterpene was zingiberene (46.71%) followed by valencene (7.61%), β-funebrene (3.09%) and selina-4(14),7(11)-diene (1.03%).

3.4 Mechanism of action

Various mechanisms have been proposed for the antidiabetic and hypoglycemic activities of medicinal plants. These include peripheral utilization of glucose, increased synthesis of hepatic glycogen by enhancement of glycogen regulatory enzyme expression in the liver, inhibition of carbohydrate metabolizing enzymes, stimulation of pancreatic insulin release, insulomimetic actions and inhibition of hepatic glucose production [55, 56].

According to Dearlove et al. [57], spices such as cinnamon, cloves, oregano, and allspice possess bioactive compounds that have (1) antiglycation properties which inhibit the formation of AGEs; (2) antioxidant activities that neutralize the

In vitro study	Result/outcome	References
[6]-Gingerol on 3 T3-L1 cells	Enhanced differentiation of 3T3-L1 preadipocytes and insulin-sensitive glucose uptake	Sekiya et al. [42]
[6]-Shogaol or [6]-gingerol on 3 T3-L1 cells	Significant inhibition of TNF-α-mediated adiponectin expression in 3T3-L1 adipocytes. [6]-Shogaol acted as a peroxisome proliferator-activated receptor (PPAR)γ agonist, while [6]-gingerol acted by suppressing TNF-α-induced JNKs signaling	Isa et al. [43]
Ethyl acetate extract of ginger on L6 myotube cell surface	Stimulated glucose uptake and GLUT4 expression in L6 myotube cell surface, reduced lipid content in 3T3 adipocyte, and inhibited protein glycation. Inhibited α-amylase (IC_{50} = 980.2 μg/mL) and α-glucosidase (IC_{50} = 180.1 μg/mL)	Rani et al., [44]
Aqueous extract of ginger at 5, 10, 20, 40 g/L incubated with (PBS), glucose + BSA for 5 weeks	Dose-dependent, antidiabetic activity through inhibition of glucose diffusion and reduced glycation	Sattar et al., [45]

Table 1.
In vitro hypoglycemic potentials of ginger and its bioactive constituents.

Human trials/dosage	Results	References
Type 2 diabetic men (40–60 years) given 3 g/day of dry ginger powder in divided doses for 30 days. Number-8 T2D, 8 placebo	Significant reduction of blood glucose, triglyceride, total cholesterol, LDL and VLDL cholesterol	Andallu et al. [48]
Randomized double-blind placebo-controlled trial, patients with Type 2 diabetes, given 2 g/day of ginger extract supplementation. Number-28 T2D, 30 Placebo	Significantly lowered levels of insulin, LDL-C, TG, HOMA index and increased the QUICKI index; no significant changes in FPG, TC, HDL-C and HbA1c; improved insulin sensitivity	Mahluji et al. [49]
Randomized controlled trial of Type 2 diabetic men between 30 and 70 years given 1.6 g/day of ginger or wheat flour capsule. Number-33 T2D, 30 Placebo	Decreased fasting blood glucose, glycosylated hemoglobin, fasting insulin, homeostasis model assessment-insulin resistance index, total cholesterol and triglyceride. No change in BMI, LDL-C, LDL-C and HDL-C	Arablou et al. [50]
Randomized controlled trial of Type 2 diabetic patients (30–70 years) given either 3 g/day ginger or cellulose microcrystalline capsules for 8 weeks. Number-40 T2D, 41Placebo	Significant reduction in fasting blood glucose and glycosylated hemoglobin; no change in BMI, fasting insulin and homeostasis model assessment-insulin resistance index	Mozaffari-Khosravi et al. [51]
Randomized, double-blind, placebo-controlled, clinical trial where Type 2 diabetic patients received 2 g/day of ginger powder supplement or lactose as placebo for 12 weeks. Number-22 T2D, 19 placebo	Significant reduction of fasting blood sugar, hemoglobin A1c, apolipoprotein B, apolipoprotein B/apolipoprotein A-I and malondialdehyde in ginger group compared to baseline and control group, while increasing apolipoprotein A-I in Type 2 diabetic patients	Khandouzi et al. [52]
Randomized controlled trial of Type 2 diabetic patients (20–60 y) T2DM given 3 g ginger or lactose capsule/day for 3 months. Number-22 T2D, 23 Placebo	Reduced fasting blood glucose, glycosylated hemoglobin, fasting insulin, homeostasis model assessment-insulin resistance index	Shidfar et al. [53]
Double-blind placebo-controlled trials of Type 2 diabetic patients were randomly allocated to 2000 mg/day of ginger or placebo for 10 weeks. Number-25 T2D, 25 placebo	Reduced serum levels of fasting blood glucose, hemoglobin A1C compared to placebo group, reduced ratio of LDL-C/HDL-C; but no significant change in serum concentrations of triglycerides, total cholesterol, LDL-C, and HDL-C	Arzatii et al. [54]

Table 2.
Clinical (human) trials of the antidiabetic potentials of ginger.

effects of ROS; and (3) anti-inflammatory potentials. Some studies associate the antidiabetic action of ginger to its bioactive principles such as gingerol and shogaol which have the capacity to enhance glucose uptake in rat's skeletal muscle cells, and promote increased expression and translocation of GLUT-4 glucose transporter to the plasma membrane of the cells thus clearing excess glucose from the serum [34].

Another mechanism proposed was the inhibition of key enzymes of carbohydrate metabolism-α-glucosidase and α-amylase by phenolic compounds (gingerols and shogaols) present in ginger [23, 45]; while other authors showed that ginger increases muscle and liver glycogen stores by enhancing peripheral utilization of glucose, thus limiting gluconeogenesis in the liver and kidney in a manner similar to insulin [37].

Son et al. [58], posits that-gingerol exerts its antidiabetic effects through multiple mechanisms that include—(1) increased glucose uptake in the absence of insulin, (2) induction of 5′ adenosine monophosphate-activated protein kinase phosphorylation, (3) promotion of glucose transporter 4 (GLUT4) translocation to plasma membrane, (4) suppression of advanced glycation end product-induced rise of ROS levels in pancreatic β-cells, (5) reduction of fasting blood glucose levels and improved glucose intolerance, (6) regulation of hepatic gene expression of enzymes involved in glucose metabolism toward decreased gluconeogenesis and glycogenolysis, while increasing glycogenesis, thereby reducing blood glucose concentrations.

4. Conclusion

This study presented an update on the antidiabetic potentials of ginger from the Zingiberaceae family. Although several in vivo and in vitro reports were available, there were relatively few clinical (human) trials. The doses and outcomes also varied; as well as the mechanism of action through which antidiabetic effects were mediated. Although these reports are indicative of the anti-diabetic or hypoglycemic potentials of ginger, the doses and outcomes also varied; most importantly, the mechanisms of action through which anti-diabetic effects are mediated were highlighted. Ginger, according to these studies, exerts its anti-diabetic effects through restorative effects on pancreatic β-cells, increasing insulin sensitivity, insulin-like action and peripheral utilization of glucose. Other mechanisms include increased synthesis of hepatic glycogen through the enhancement of glycogen regulatory enzyme expression in the liver, inhibition of carbohydrate metabolizing enzymes, stimulation of pancreatic insulin release, and inhibition of hepatic glucose production. However, further studies, especially in humans are therefore needed and the oral safety of the various extracts under prolonged usage must be confirmed, more so, since ginger is one of the spices generally regarded as safe.

Acknowledgements

The authors acknowledge the support of Govan Mbeki Research and Development Centre, University of Fort Hare.

Conflict of interest

The authors declare no conflict of interest.

Author details

Gloria Aderonke Otunola* and Anthony Jide Afolayan
Medicinal Plants and Economic Development (MPED) Research Centre, Department of Botany, University of Fort Hare, Alice, South Africa

*Address all correspondence to: gotunola@ufh.ac.za

References

[1] Ogurtsova K, da Rocha Fernandes JD, Huang Y, Linnenkamp U, Guariguata L, Cho NH, et al. Global estimates for the prevalence of diabetes for 2015 and 2040. IDF Diabetes Atlas;**2017**(128):40-50

[2] Bi X, Lim J, Henry CJ. Spices in the management of diabetes mellitus. Food Chemistry. 2017;**217**:281-293

[3] Hui, Hongxiang, Xiaoning Zhao, and Riccardo Perfetti. Structure and function studies of glucagon-like peptide-1 (GLP-1): The designing of a novel pharmacological agent for the treatment of diabetes. Diabetes/Metabolism Research and Reviews. 2005;**21**(4):313-331

[4] Bathaie SZ, Mokarizade N, Shirali S. An overview of the mechanisms of plant ingredients in the treatment of diabetes mellitus. Journal of Medicinal Plants. 2012;**4**(44):1-24

[5] Kooti W, Farokhipour M, Asadzadeh Z, Ashtary-Larky D, Asadi-Samani M. The role of medicinal plants in the treatment of diabetes: A systematic review. Electronic Physician. 2016;**8**(1):1832

[6] Kumari KD, Suresh KP, Samarasinghe K, Handunnetti SM, Samaranayake TSP. Evaluation of a traditional Sri Lankan herbal beverage (water extract of dried flowers of *Aegle marmelos*, Bael fruit) in type II diabetic patients. Journal of Diabetes and Metabolism. 2013;**4**(6)

[7] Haque N, Salma U, Nurunnabi TR, Uddin MMJ, Jahangir FK, Islam SMZ, et al. Management of type 2 diabetes mellitus by lifestyle, diet and medicinal plants. Pakistan Journal of Biological Sciences. 2011;**14**:13-24

[8] Viuda-Martos M, Ruiz-Navajas Y, Fernández-López J, Pérez-Álvarez JA. Spices as functional foods. Critical Reviews in Food Science and Nutrition. 2010;**51**(1):13-28

[9] Afzal M, Al-Hadidi D, Menon M, Pesek J, Dhabi MSG. An ethnomedicinal, chemical and pharmacological review. Drug Metabolism and Drug Interactions. 2001;**18**:159-190

[10] Bode AM, Dong Z. The amazing and mighty ginger. In: Benzie IFF, Wachtel-Galor S, editors. Herbal Medicine: Biomolecular and Clinical Aspects. 2nd ed. Boca Raton, FL: CRC Press/Taylor & Francis; 2011. Available from: https://www.ncbi.nlm.nih.gov/books/NBK92775/

[11] Grzanna R, Lindmark L, Frondoza CG. Ginger—An herbal medicinal product with broad anti-inflammatory actions. Journal of Medicinal Food. 2005;**8**(2):125-132

[12] Ali BH, Blunden G, Tanira MO, Nemmar A. Some phytochemical, pharmacological and toxicological properties of ginger (*Zingiber officinale* Roscoe): A review of recent research. Food and Chemical Toxicology. 2008;**46**(2):409-420

[13] Ghayur MN, Gilani AH. Pharmacological basis for the medicinal use of ginger in gastrointestinal disorders. Digestive Diseases and Sciences. 2005;**50**:1889-1897. DOI: 10.1007/s10620-005-2957-2

[14] Nicoll R, Henein MY. Ginger (*Zingiber officinale* Roscoe): A hot remedy for cardiovascular disease? International Journal of Cardiology. 2009;**131**:408-409

[15] Wu KL, Rayner CK, Chuah SK. Effects of ginger on gastric emptying and motility in healthy humans. European Journal of Gastroenterology & Hepatology. 2008;**20**(5):436-440

[16] Qian QH, Yue W, Wang YX, Yang ZH, Liu ZT, Chen WH. Gingerol inhibits cisplatininduced vomiting by down regulating 5-hydroxytryptamine, dopamine and substance P expression in minks. Archives of Pharmacal Research. 2009;**32**(4):565-573

[17] Ojewole JAO. Analgesic, anti inflammatory and hypoglycaemic effects of ethanol extract of *Zingiber officinale* (Roscoe) rhizomes (Zingiberaceae) in mice and rats. Phytotherapy Research. 2006;**20**(9):764-772

[18] Elkhishin IA, Ibrahim AA. A study of the cardiovascular toxic effects of *Zingiber officinale* (ginger) in adult male albino rats and its possible mechanisms of action. Mansoura Journal of Forensic Medicine and Clinical Toxicology. 2009;**17**(2):109-127

[19] Singh A, Sanjiv D, Jaswinder S, Shankar K. Experimental advances in pharmacology of gingerol and analogues. Pharmacy Global: International Journal of Comprehensive Pharmacy. 2010;**2**(4)

[20] Salim KS. Hypoglycemic property of ginger and green tea and their possible mechanisms in diabetes mellitus. Open Conference Proceedings Journal. 2014;**5**:13-19

[21] Yiming L, Van HT, Colin CD, Basil DR. Preventive and protective properties of *Zingiber officinale* (ginger) in diabetes mellitus, diabetic complications, and associated lipid and other metabolic disorders: A brief review. Evidence-based Complementary and Alternative Medicine. 2012:10. Article ID 516870

[22] Kalejaiye OF, Iwalewa EO, Omobuwajo OR, Oyedapo OO. Hypoglycaemic effects of Nigerian *Zingiber officinale* rhizome on experimental diabetic rats. Nigerian Journal of Natural Products and Medicine. 2002;**6**(1):33-35

[23] Nammi S, Satyanarayana S, Basil DR. Protective effects of ethanolic extract of *Zingiber officinale* rhizome on the development of metabolic syndrome in high-fat diet-fed rats. Basic & Clinical Pharmacology & Toxicology. 2009;**104**(5):366-373

[24] Abdulrazaq NB, Maung MC, Ni NW, Rahela Z, Mohammad TR. Beneficial effects of ginger (*Zingiber officinale*) on carbohydrate metabolism in streptozotocin-induced diabetic rats. British Journal of Nutrition. 2012;**108**(7):1194-1201

[25] Jafri SA, Sohail A, Muhammad Q. Hypoglycemic effect of ginger (*Zingiber officinale*) in alloxan induced diabetic rats (*Rattus norvagicus*). Pakistan Veterinary Journal. 2011;**31**(2):160-162

[26] Morakinyo AO, Akindele AJ, Ahmed Z. Modulation of antioxidant enzymes and inflammatory cytokines: Possible mechanism of anti-diabetic effect of ginger extracts. African Journal of Biomedical Research. 2011;**14**(3):195-202

[27] Lindstedt I. Ginger and diabetes: A mini-review. Archives of General Internal Medicine. 2018;**2**(2):29-33

[28] www.amazon.com/Culinary-Ginger-Zingiber-Officinalecooking/dp/B01NAJQ6C6 [Accessed: 18 July 2018]

[29] Lamuchi-Deli N, Mohammad A, Hossein B-R, Ghorban M. Effects of the hydroalcoholic extract of *Zingiber officinale* on arginase I activity and expression in the retina of streptozotocin-induced diabetic rats. International Journal of Endocrinology and Metabolism. 2017;**15**(2)

[30] de La Heras N, Munoz VM, Fernandez MB, Ballesteros S, Farre LA, Roso RB, et al. Molecular factors involved in the hypolipidemic- and insulin-sensitizing effects of a ginger (*Zingiber officinale* Roscoe) extract

in rats fed a high-fat diet. Applied Physiology, Nutrition, and Metabolism. 2017;**42**:209-215

[31] Wei CK, Tsai YH, Korinek M, Hung PH, El-Shazly M, Cheng YB, et al. 6-Paradol and 6-shogaol, the pungent compounds of ginger, promote glucose utilization in adipocytes and myotubes, and 6-paradol reduces blood glucose in high-fat diet-fed mice. International Journal of Molecular Sciences. 2017;**18**:168

[32] Al-Qudah MMA, Moawiya AH, El-Qudah JMF. The effects of aqueous ginger extract on pancreas histology and on blood glucose in normal and alloxan monohydrate-induced diabetic rats. Biomedical Research. 2016;**27**(2)

[33] Oludoyin AP, Adegoke SR. Effect of ginger (*Zingiber officinale*) extracts on blood glucose in normal and streptozotocin-induced diabetic rats. International Journal of Clinical Nutrition. 2014;**2**:32-35

[34] Anfenan MLK. Evaluation of nutritional and antidiabetic activity of different forms of ginger in rats. Middle-East Journal of Scientific Research. 2014;**21**:56-62

[35] Al-Noory AS, Amreen AN, Hymoor S. Antihyperlipidemic effects of ginger extracts in alloxan-induced diabetes and propylthiouracil-induced hypothyroidism in (rats). Pharmacognosy Resarch. 2013;**5**:157-161

[36] Abdulrazaq N, Cho M, Win N, Zaman R, Rahman M. Beneficial effects of ginger (*Zingiber officinale*) on carbohydrate metabolism in streptozotocin-induced diabetic rats. British Journal of Nutrition. 2012;**108**(7):1194-1201. DOI: 10.1017/S0007114511006635

[37] Iranloye BO, Arikawe AP, Rotimi G, Sogbade AO. Anti-diabetic and anti-oxidant effects of *Zingiber officinale* on alloxan-induced and insulin-resistant diabetic male rats. Nigerian Journal of Physiological Sciences. 2011;**26**(1)

[38] Akhani SP, Vishwakarma SL, Goyal RK. Anti-diabetic activity of *Zingiber officinale* in streptozotocin-induced type I diabetic rats. The Journal of Pharmacy and Pharmacology. 2004;**56**:101-105

[39] Al-Amin ZM, Thomson M, Al-Qattan KK, Peltonen-Shalaby R, Ali M. Anti-diabetic and hypolipidaemic properties of ginger (*Zingiber officinale*) in streptozotocin-induced diabetic rats. British Journal of Nutrition. 2006;**96**(4):660-666

[40] Bhandari U, Kanojia R, Pillai K. Effect of ethanolic extract of *Zingiber officinale* on dyslipidaemia in diabetic rats. Journal of Ethnopharmacology. 2005;**97**:227-230. DOI: 10.1016/j.jep.2004.11.011

[41] Otunola GA, Afolayan AJ. Antidiabetic effect of combined spices of *Allium sativum, Zingiber officinale* and *Capsicum frutescens* in alloxan-induced diabetic rats. Frontiers in Life Science. 2015;**8**(4):314-323

[42] Sekiya K, Ohtani A, Kusano S. Enhancement of insulin sensitivity in adipocytes by ginger. Bio Factors. 2004;**22**(1-4):153-156

[43] Isa Y, Miyakawa Y, Yanagisawa M, et al. [6]-Shogaol and [6]-gingerol, the pungent of ginger, inhibit TNF-alpha mediated downregulation of adiponectin expression via different mechanisms in 3T3-L1 adipocytes. Biochemical and Biophysical Research Communications. 2008;**373**(3):429-434

[44] Rani MP, Krishna MS, Padmakumari KP, Raghu KG, Sundaresan A. *Zingiber officinale* extract exhibits antidiabetic potential via modulating glucose uptake, protein glycation and inhibiting adipocyte

differentiation: An in vitro study. Journal of the Science of Food and Agriculture. 2012;**92**(9):1948-1955

[45] Sattar NA, Hussain F, Iqbal T, Sheikh MA. Determination of in vitro antidiabetic effects of *Zingiber officinale* Roscoe. Brazilian Journal of Pharmaceutical Sciences. 2012;**48**(4):601-607

[46] Andallu B, Radhika B, Suryakantham V. Effect of aswagandha, ginger and mulberry on hyperglycemia and hyperlipidemia. Plant Foods for Human Nutrition. 2003;**58**(3):1-7

[47] Mahluji S, Attari VE, Mobasseri M, Payahoo L, Ostadrahimi A, Golzari SEJ. Effects of ginger (*Zingiber officinale*) on plasma glucose level, HbA1c and insulin sensitivity in type 2 diabetic patients. International Journal of Food Sciences and Nutrition. 2013;**64**(6):682-686. DOI: 10.3109/09637486.2013.775223

[48] Arablou T, Aryaeian N, Valizadeh M, Sharifi F, Hosseini A, Djalali M. The effect of ginger consumption on glycemic status, lipid profile and some inflammatory markers in patients with type 2 diabetes mellitus. International Journal of Food Sciences and Nutrition. 2014;**65**(4):515-520

[49] Mozaffari-Khosravi H, Talaei B, Jalali BA, Najarzadeh A, Mozayan MR. The effect of ginger powder supplementation on insulin resistance and glycemic indices in patients with type 2 diabetes: A randomized, double-blind, placebo-controlled trial. Complementary Therapies in Medicine. 2014;**22**(1):9-16

[50] Khandouzi N, Farzad S, Asadollah R, Tayebeh R, Payam H, Mohsen MT. The effects of ginger on fasting blood sugar, hemoglobin A1c, apolipoprotein B, apolipoprotein AI and malondialdehyde in type 2 diabetic patients. Iranian Journal of Pharmacy Research. 2015;**14**(1):131

[51] Shidfar F, Rajab A, Rahideh T, Khandouzi N, Hosseini S, Shidfar S. The effect of ginger (*Zingiber officinale*) on glycemic markers in patients with type 2 diabetes. Journal of Complementary and Integrative Medicine. 2015;**12**(2):165-170

[52] Arzati MM, Honarvar NM, Saedisomeolia A, Anvari S, Effatpanah M, Arzati RM, et al. The effects of ginger on fasting blood sugar, hemoglobin A1c, and lipid profiles in patients with type 2 diabetes. International Journal of Endocrinology and Metabolism. 2017;**15**(4):e57927. DOI: 10.5812/ijem.57927

[53] Koch W, Kukula-Koch W, Marzec Z, Kasperek E, Wyszogrodzka-Koma L, Szwerc W, et al. Application of chromatographic and spectroscopic methods towards the quality assessment of ginger (*Zingiber officinale*) rhizomes from ecological plantations. International Journal of Molecular Sciences. 2017;**18**(2):452

[54] Sharma PK, Singh V, Ali M. Chemical composition and antimicrobial activity of fresh rhizome essential oil of *Zingiber officinale* Roscoe. Pharmacognosy Journal. 2016;**8**(3):185-190

[55] Bnouham M, Ziyyat A, Mekhfi H, Tahri A, Legssyer A. Medicinal plants with potential antidiabetic activity—A review of ten years of herbal medicine research (1990-2000). International Journal of Diabetes and Metabolism. 2006;**14**:1-25

[56] Yatoo MI, Saxena A, Gopalakrishnan A, Alagawany M, Dhama K. Promising antidiabetic drugs, medicinal plants and herbs: An update. International Journal de Pharmacologie. 2017;**13**:732-745

[57] Dearlove RP, Greenspan P, Hartle DK, Swanson RB, Hargrove JL. Inhibition of protein glycation

by extracts of culinary herbs and spices. Journal of Medicinal Food. 2008;**11**(2):275-281

[58] Son MJ, Miura Y, Yagasaki K. Mechanisms for antidiabetic effect of gingerol in cultured cells and obese diabetic model mice. Cytotechnology. 2015;**67**:641

Chapter 9

Pharmacological Potentials of Ginger

Fatai Oladunni Balogun, Esther Tayo AdeyeOluwa and Anofi Omotayo Tom Ashafa

Abstract

Zingiber officinale, belonging to the family Zingiberaceae, is a popular spice and herb used as delicacy and to manage numerous diseases such as diabetes, hypertension, cancer, ulcer, diarrhea, cold, cough, spasm, vomiting, etc. in folk medicine from China, India, and Arabia Peninsula to other continents of the world including Africa (Nigeria, Egypt, and so on). Though this review is aimed at summarizing the pharmacological potentials of this well-endowed spice, interestingly, we found out that these reported ethnobotanical uses are attributed to a number of inherent chemical constituents including gingerol, 6-, 8-, 10-gingerol, 6-shogaol, 6-hydroshogaol, oleoresin, etc., eliciting various pharmacological effects, not limited to antioxidant, antitumor/anticancer, anti-inflammatory, antihyperglycemic, antihypertensive, anticholesterolemic, antibiotic/antimicrobial, neuroprotective, antiulcer/gastroprotective, antiemetic, hepatoprotective, and antiplatelet aggregation, safety profiles established through a number of studies (in vitro, in vivo, and cell lines), though some of these potentials are yet to be explored. Sadly, even few of these established effects are yet to be experimented in clinical trials, and only until these are intensified would there be prospect toward drug development for preventive and curative treatments. In conclusion, we are able to highlight and sum up the therapeutic implications of ginger and its related derivatives in the management of ailments confronting humanity.

Keywords: ginger, spice, pharmacological potentials, gingerol, 6-, 8-, 10-gingerol, 6-shogaol, 6-hydroshogaol, oleoresin

1. Introduction

Ginger (*Zingiber officinale* Roscoe) is a well-known herbal spice believed to have originated from either India [1] or Southeast Asia [2]. It is a sterile plant, thus reproduced by rhizomes, not by seeds [3], and grows well in tropical and subtropical regions of the world [4]. It is used for culinary purposes, as a seasoning or condiment and as a therapeutic agent [5]. It is known to be an effective spasmolytic, antipyretic, antiemetic, antioxidant, antiulcer, analgesic, hypotensive, antidiabetic, and anti-inflammatory agent [6, 7] containing scented essential oils and spicy oleoresins [8]. Ginger has long been in use therapeutically and currently still validated as a potent medicinal spice for the treatment of various ailments. Indigenously, it has been used against colds [9], sore throats [10], and *Staphylococcus aureus* [11] and tested effectively against cancer cells [12]. Ginger can be used as a dietary supplement and as additives in the production of various snacks and merchantable products [13]. Additionally, it is considered a safe herbal drug [14], as the spices have been categorized to be generally regarded as safe: "GRAS."

2. Botanical description, occurrence, and distribution

Zingiber officinale (Roscoe), ginger of the family Zingiberaceae, is an herbaceous (available as rhizomes) perennial plant growing as tall as 90 cm. The leaves, lanceolate, appear to be simple, alternate, distichous, narrow, long possessing sheathing bases with 2–3 cm broad, while the rhizomes (7–15 cm long and 1–1.5 cm broad) are aromatic, thick lobed with pale yellow coloration. The flowers are small, have calyx that are lofty, have sepals very united, are three toothed, and split open on a side with three subequal corolla forming an oblong to lanceolate connate segment with green coloration [15, 16]. Ginger give rise to numerous lateral clump shoot which on maturation appeared dry. Ginger originate from Southeast Asia predominately in India but now well distributed or cultivated in China, Bangladesh, Australia, and Nigeria [17].

3. Ethnobotanical uses

Ginger had been used medicinally since time immemorial with documented use from Sanskrit, Chinese, Greek, Arabic, and Roman ethnomedicine book. However, in the ninth century, Europe recognized the indigenous use of this wonderful spice, and England followed suit in the tenth century. Ginger is used in folkloric medicine for indigestion, high blood pressure, arthritis, intestinal and throat infections, vomiting, nausea, lung diseases, cold, cough, pain, swellings, etc. [15, 17, 18]. Other nutritional uses are found in condiment, beer, wine, and so on [18].

4. Phytochemistry of ginger

Ginger, a spice of diverse health benefits, has been found to be rich in nonnutritive and biologically active compounds known as phytochemicals [19, 20], which have been linked to its health functions. The nutritional and therapeutic values have been recognized in its nutraceutical benefits linked to the presence of certain phytochemicals contained in it. The use of ginger as a nutraceutical agent is not only attributed to its health-augmenting benefits but also to its availability, affordability, and safety.

More than 400 compounds have been found in the chemical analyses of ginger [21]. These compounds includes alkaloids; saponins; flavonoids; steroids; tannins; carbohydrates; glycosides; proteins; amino acids; dietary fiber; ash; phytosterols; vitamins A, B, and C; minerals; and terpenoids [22–24] while detected to be devoid of acid compounds and reducing sugars [23].

The main components of the ginger rhizome are in the order carbohydrates, lipids, terpenes, and phenolic compounds [25]. The terpenes and the phenolic compounds make up the two foremost classes of phytochemicals in ginger [26]. Phenolic compounds of ginger are also referred to as its nonvolatile components, which have been incriminated in its pharmacological activity. They consist of gingerols and its 6, 8, and 10 derivatives and the corresponding series of homologous shogaol and zingerone, obtained from heat or alkali treated gingerols [26]. Shogaol, paradol, and gingerols have been depicted to be responsible for the pungent taste and smell of ginger [1, 27]. The terpene components of ginger, sesquiterpenes and monoterpenes, are believed to be the volatile fractions [27]. The sesquiterpenes are thought to be a major contributor to the savor of ginger, while the monoterpenes are referred to as the most abundant terpenes in fresh ginger oil [24]. The main sesquiterpenes, zingiberene and β-bisabolene, are responsible for its aromatic scent, while others include α-farnesene, β-sesquiphellandrene, and α-curcumene [21].

Phenolic compounds of ginger are majorly derived from fresh ginger rhizomes, while the terpenes are derived from distillation of ginger oils [26] although their quantity has been found to vary depending on the region of germination. This may be dependent on climate or edaphic conditions as well as genetic variations [28]. The pungent compounds (gingerols, methyl gingerols, shogaols, paradol, and gingerdiones), volatile oil, and other compounds extracted by means of ethanol or acetone constitute the oleoresin [29, 30]. Volatile oils are about 1–4%, lipids about 6–8%, proteins about 9%, and carbohydrates about 50–80% [28] while geraniol is the major essential oil derived in ginger [8].

Zingerone, geraniol, gingerols, shogaols, gingerdiols, gingerdiones, and dehydrogingerdiones have been reported to have antioxidant activity; 6-, 8-, and 10-gingerol and 6-gingerdiol possessed antifungal activity. While 6-gingerol had established antidiabetic and reno-protective activities, zingerone, 6-shogaol, 6-gingerol (anticancer, anti-obesity, and gastroprotective activities), and gingerol and its pungent derivatives (anti-inflammatory activity), 6-shogaol (analgesic, neuroprotective, and strong gastroprotective activities), 6-gingerol, and 6-shogaol, acted against platelet aggregation; 10-gingerol had larvicidal activity; and 6-, 8-, 10-gingerol possessed inotropic activity [24].

5. Pharmacological potentials

The review from most countries of the world such as Egypt [20], Korea [17], Pakistan [15], India [16, 31, 32], Oman [5], Brazil [33], Canada [34] etc. had established the pharmacological potentials of this popular plant, *Zingiber officinale* (Roscoe), used most times as spice. Additionally, while some reports centers on the action of ginger, others point to the effect of its active components as they target specific diseases including but not limited to diabetes [35], inflammation [25], cancer [22], emetics [36], nausea and vomiting [37] and so on. Thus, the pharmacological potentials (antioxidant, anticancer, antitumor, anti-inflammatory, antihyperglycemic, antihypertensive, anticholesterolemic, antimicrobial, neuroprotective, antiulcer, antiemetic, hepatoprotective) and toxicity profiles of ginger as submitted in these reports are presented one after the other below.

5.1 Antioxidant

The overproduction of free radicals (ROS) in situation where the antioxidant defense mechanism is compromised results into a state of oxidative stress. In order to overcome the excessive free radical (FR) generation and oxidative stress, antioxidants play an important role. Numerous medicinal plants (MPs) and/or their constituents have established their prominence in preventing the onset of diseases particularly those triggered by FR. Ginger, a good example of MPs with excellent antioxidative effect, has been found to exert this action by lowering peroxidation of lipid such as the inhibition of ascorbate/ferrous complex in rat liver microsomes as cited by Rahmani et al. [20] and Mele [17] in the report of Reddy and Lokesh [38] using a concentration of 150 mM (**Table 1**). Ginger or its derivatives (extracts, compounds, or active components) and gingerol are found to have good scavenging effect against superoxide anion and hydroxyl radicals [63–65]. In fact, further

Ginger/derivatives	Potentials	Assay(s) employed	Type of study	Concentration(s) tested	Extracts (if any)	Country (where the report is published)	References
Ginger	Antioxidant	Lipid peroxidation	In vivo (rats)	150 mM	NI	India	Reddy and Lokesh, [38]
Ginger extracts	Hepatoprotective	Thioacetamide-induced (200 mg/kg i.p)	In vivo (rats)	250, 500 mg/kg	Ethanol	Malaysia	Bardi et al., [39]
Ginger extracts	Antiproliferative	Hep G2 cells	Cell lines	NI	Ethanol	Malaysia	Bardi et al., [39]
6-gingerol	Antioxidant, anti-inflammatory	UVB-induced intracellular reactive oxygen species levels	In vitro and in vivo (mice)	NI (200 μL)	Acetone	South Korea	Kim et al., [40]
Ginger	Antioxidant, antidiabetic	MDA, FRAP streptozotocin-induced	In vitro and in vivo (rats)	5% ginger in daily foods	NI	Iran	Afshari et al., [41]
Oleoresin, 6-gingerol, 8-gingerol, 10-gingerol, 6-shogaol, 6-hydroshogaol	Antioxidant, antimicrobial, anti-inflammatory	Nitric oxide, ABTS, DPPH, Disc diffusion	In vitro	NI	NI	India, Algeria	Dugasani et al., [42]; Li et al., [43]; Bellik, [44]
Ginger	Anticancer (prostrate and liver)	0.1% ethionine-induced	In vitro and in vivo	100 mg/kg	NI	USA, Malaysia	Karna et al., [45]; Habib et al., [46]
Ginger extracts	Anticancer (pancreas)	Panc-1 cells	Cell lines, in vivo	NI	Ethanol	Japan	Akimoto et al., [47]
	Anti-inflammatory, analgesic, hypoglycemic, safety profile	Hind paw	In vitro and in vivo	50–800 mg/kg	Ethanol	South Africa	Ojewole, [48]

Ginger/derivatives	Potentials	Assay(s) employed	Type of study	Concentration(s) tested	Extracts (if any)	Country (where the report is published)	References
Ginger oil	Antiarthritis, anti-inflammatory	Hind paw	In vivo	33 mk/kg	NI	Malaysia	Sharma et al., [49]
Ginger	Antidiabetic	Streptozotocin -induced	In vivo	100, 300, 500 mg/kg bw	Aqueous	Malaysia	Abdul Razaq et al., [50]
Ginger	Antidiabetic	Standard spectrophotometric methods (glycation inhibition, glucose diffusion)	In vitro	5, 10, 20, 40 g/L	Aqueous	Pakistan	Sattar et al., [51]
6-gingerol	Antidiabetic	Streptozotocin-induced	In vivo	25, 50 mg/kg bw	Aqueous	Malaysia	Sukalingam et al., [52]
Ginger, 6-, 8-, and 10-gingerol, while 6-shogaol	Antihypertensive	Pentothal -induced	In vivo	3–10 mg/kg	Aqueous	Pakistan	Ghayur et al., [53]
Ginger	Anticholesterolemic	NI	In vivo	50, 500 mg/kg	Aqueous	Kuwait	Thompson et al., [54]
Ginger	Antibacterial	Disc diffusion	In vitro	0.125, 0.25, 0.5, 1.0% 35.25, 75, 250, 500 mg/ml 20, 40, 60, 80, 100 g/ml	Ethanol, ethyl acetate, n-hexane aqueous, ethanol aqueous, ethanol	Nigeria Nigeria	Malu et al., [55] Ekwenye and Elegalam, [56] Sebiomo et al., [57]
Oleoresin, essential oil	Antifungi and antibacterial (antimicrobial)	Disc diffusion	In vitro	NI (3 uL)	NI	Algeria	Bellik, [44]

Ginger/derivatives	Potentials	Assay(s) employed	Type of study	Concentration(s) tested	Extracts (if any)	Country (where the report is published)	References
Ginger	Neuroprotective	Monosodium glutamate-induced	In vivo	100 mg/kg	Aqueous	Saudi Arabia	Waggas, [58]
Ginger 6-gingerol	Gastroprotective	HCl-ethanol induced	In vivo	1000 mg/kg 100 mg/kg	Acetone	Japan	Johji et al., [59]
6-gingerol, 6-shogaol	Gastric suppression	Hexobarbital induced	In situ	1.75–3.5 mg/kg (i.v), 70–140 mg/kg (oral)	NI	Japan	Suwekawa et al., [60, 61]
Ginger	Hepatoprotective	Carbon tetrachloride-induced Acetaminophen-induced	In vivo In vivo	100 mg/kg (singly), 50 mg/kg (combination with 100 mg/kg curcumin) 200, 400 mg/kg	NI Aqueous ethanol	Egypt India	Abdullah et al., 2016 Ajith et al., [62]

NI: Not indicated

Table 1.
Pharmacological potentials of ginger and its derivatives.

reports indicated that upon further heating (ginger), this activity remained unaffected [66]. Furthermore, it diminishes the ultraviolet B (UVB)-induced intracellular reactive oxygen species (ROS) and cyclooxygenase (COX)-2 in in vitro and in vivo studies [40]. Other derivatives of ginger such as oleoresin, 6-shogaol, 6-dehydroshogaol, 1-dehydro-6-gingerdione, 6-gingerol, 8-gingerol, 10-gingerol, and essential oil possess pharmacological activities such as antioxidant, antimicrobial, etc., against 2,2-azino-bis-3-ethylbenzothiazoline-6-sulfonic acid (ABTS), 1,1-diphenyl-2-picrylhydrazyl (DPPH), hydroxyl radical, and microbial strains such as *Bacillus subtilis*, *Bacillus cereus*, *Staphylococcus aureus*, *Escherichia coli*, *Candida albicans*, *Penicillium* spp., and *Aspergillus niger* [42–44].

5.2 Antitumor/anticancer

Cancer is one of the noncommunicable diseases with great negative impact on global population. It is caused by persistent increase in abnormal human body cells leading to the formation of tumors (of malignant cells) with the possibility to be metastatic [67]. The continuous multiplication of these cells is sometimes associated to influence oxidative stress. A number of treatment (chemotherapy, radiotherapy, synthetic drugs, etc.) are currently available; however, they come with one or several side effects (nausea, hair fall), hence, the need for alternative form of treatment or therapy particularly from MPs. In recent times, quite a number of plant species had found their relevance in the prevention and treatment of cancer, and efforts of researchers to continually develop new moieties are overwhelming. Ginger is a great example of such MPs with excellent prophylactic and curative anticancer properties. Although it must be noted that these effects are not available for all cancer types, several reports on ginger and its derivative (gingerol) have established numerous effects on different types of cancer (lung, colon, ovarian, prostrate, etc.) in a study conducted in the United States by Karna et al. [45] however, daily oral administration of ginger at a concentration of 100 mg/kg body weight (bw) inhibited PC-3 xenograft growth, indicating its effect against prostate cancer in vitro and in vivo. Additionally, the same concentration in another study [46] reduced the increased activity of tumor necrosis factor-alpha (TNF-α) due to the blockage of rat's liver cancer. Its effect on Panc-1 cells and other cell lines in an in vitro and animal model had been established against cancer of the pancreas, while combining the spice with other spices such as garlic and turmeric provided effectiveness against breast cancer [68]. In line with the aforementioned effects, derivatives of ginger, e.g., 6-shogaol, 8-shogaol, 10-shogaols, 6-gingerol, 6-paradol, and zingerone in several studies had also exhibited activities against different form of cancer including lung, colon, colorectal, ovarian, prostrate as cited by Rahmani et al. [20] and Gunathilake and Rupasinghe [34] from numerous studies. Interestingly, ginger was also reported to hinder tumor growth achieved through different molecular mechanism such as upregulation of suppressor gene, apoptosis, induction, and inactivation of vascular endothelial growth factor (VEGF) (molecular pathways), a tumor angiogenic factor that triggers tumor development and progression [20].

5.3 Anti-inflammatory

Inflammation is a response (defense) felt by the body to dangerous stimuli such as injury to tissues or allergens. However, when these responses are beyond normal, it manifest into arrays of derangements including but not limited to allergies, cancer, autoimmune disorder, metabolic syndrome, and cardiovascular diseases [69]. Interestingly, there are reports of relationship between oxidative stress-triggered FR and inflammation. The use of nonsteroidal anti-inflammatory drugs (NSAIDs) is employed to ameliorate acute and chronic types of inflammation. NSAIDs exhibit this action by inhibiting the enzyme (cyclooxygenase, COX 1 and 2, and/or lipoxygenase, 5, 10, 15) involved in the breakdown of arachidonic acid to prostaglandins. Unfortunately, there are numerous side effects emanating from the use of NSAIDs, hence the search for alternative form of treatment with minimal or no side effects in natural products. Intriguingly, numerous MPs have shown to be effective against inflammatory diseases. Ginger, an example of such MPs including its derivatives, has been reported to possess anti-inflammatory potentials [17] in vitro and in vivo studies [34]. Ojewole [48] submitted the analgesic, anti-inflammatory, hypoglycemic, and safety effect of ginger extract at a dose range of 50–800 mg/kg bw (**Table 1**). Thirty-three mg/kg bw ginger oil given to rats also alleviated acute and chronic arthritis [49]. Interestingly, ginger exhibits its anti-inflammatory activity in other solvents aside water (used in folkloric medicine), as the reports of Rani et al. [70] corroborate this when ethyl acetate-extracted ginger revealed the best anti-inflammatory effect better than water, methanol (polar solvents), and hexane (nonpolar) against cyclooxygenase and lipoxygenase known as anti-inflammatory enzymes as cited by Gunathilake and Rupasinghe [34] and Mele [17] from various reports. Additionally, ginger plays a very good role in regulating the release of mediators (nitric oxide, prostaglandins), cytokines, TNF, and interleukin (IL)-1, IL-8, via several biochemical pathways attributed to inflammation, etc. [17, 20, 25, 33, 34].

5.4 Antihyperglycemic

Diabetes mellitus (DM) is one of the noncommunicable diseases with major prevalence globally. It is an endocrine disorder or metabolic derangement characterized by hyperglycemia (elevated level of glucose in the blood) due to insufficient or ineffective insulin arising from abnormalities in carbohydrate, lipid, and protein. The treatment or management of DM could be non-pharmacological (exercise, dietary regimen) or pharmacological which entails the use of oral hypoglycemic agents (OHAs) such as sulphonyl ureas, biguanides, and so on. However, the use of these chemicals or synthetic agents is prone to side effects (obesity), unavailability, and unaffordability, hence the dire need for alternative form of treatment with little or no side effects. Surprisingly, these qualities are now found in medicinal plants. In fact, the World Health Organization in a number of their technical reports advocated and encouraged the use of MPs for diabetic control and management. It is interesting to note that quite avalanches of MPs have found their relevance as antidote to curing diabetes [71] and some of its related complications. Ginger is one of such MPs traditionally used to salvage diabetes. In fact, numerous reports are available in the literature [17, 34, 35] establishing the potential of this spice in in vitro and in vivo studies. A similar example is the report of Ojewole [48] as submitted previously in a section (above) of this report. Similarly, 500 mg/kg bw of its aqueous extract lowers plasma sugar level following streptozotocin induction [50] in animal model and in vitro [51]. Since there is a report of correlation between oxidative stress and DM [20] as well as other complications of DM such as hyperlipidemia, hypercholesterolemia, retinopathy, and neuropathy, various publications had revealed the potentials of ginger and its derivatives against these complications as cited by Gunathilake and Rupasinghe [34].

5.5 Antihypertensive

Hypertension, a silent killer (because it shows no symptoms), is characterized by continuous increase in blood pressure in the arteries of a person. It occurs when the systolic and diastolic blood pressures rise above 140/90 mmHg, respectively. Findings revealed excessive salt intake, smoking, alcohol consumption, narrowing of the kidney, and use of birth control pills as some of the causes of hypertension, a risk factor to many cardiovascular diseases (CVD). Like diabetes, the treatment option may be non-pharmacological (lifestyle modification, etc.) or pharmacological involving the use of synthetic moieties such as diuretics, beta blockers (atenolol), angiotensin-converting enzyme inhibitor (Lisinopril), calcium channel blockers, etc. [72]. However, sadly too, all these antihypertensive agents bring about grievous adverse effects such as angioedema, dry cough, weakness, headaches, etc.; thus, there is need for a substitute form of therapy for sufferers of high blood pressure (HBP). Herbal products from MPs have come very handy in the fight geared toward treating HBP, and a notable example of such plant is ginger. In a study involving rats and guinea pigs, extracts of ginger at concentration range of 0.3–3 mg/kg lower the arterial blood pressure of these animals [34]. Additionally, similar study using ginger aqueous extract and its derivatives revealed similar action [53] (**Table 1**). The activity of ginger as antihypertensive agent was also corroborated in a study [73] involving human subject when twice daily intake of 10 g of the spice reduced the arterial blood pressure to 94.80 mmHg after 2 months. It is worthy of mention that the mechanism of the established action of this spice was through the stimulation of muscarinic receptors and calcium channel blockage.

5.6 Anticholesterolemic

Cholesterol is a constituent of the plasma membranes (eukaryotic) representing sterols [74], needed for growth and development of higher organism. Hypercholesterolemia occurs when there is an elevated level of cholesterol though it suffices to say that there are good (high-density lipoprotein cholesterol (HDL-c)) and bad (low-density lipoprotein cholesterol (LDL-c)) cholesterol. Hypercholesterolemia is a risk factor to many diseases including CVD, atherosclerosis, myocardial infarction (MI), etc. [75] and there are reports of high level of cholesterol in the blood on the influence of excessive production of FRs [76–78]. The use of herbal medicines or MPs for therapeutic/curative or preventive measures against diseases is an age-long tradition [79]. Ginger is one of such herbal medicine with cholesterol-lowering properties. In a study by Thomson et al. [80] as cited by Gunathilake and Rupasinghe [34], oral administration of 500 mg/kg bw of aqueous extract brought down elevated level of cholesterol in Wistar rats. Another study using mice revealed a 29% reduction in the cholesterol level and other lipid profiles on the administration of 250 μmkg ethanolic extract studied on rabbits [81] and rats (100, 400 mg/kg bw) in a high-fat diet-fed rodents [82]. In the same vein, a study on human subject revealed a positive coadministration of atorvastatin (low dose) and ginger reducing cholesterol level in the blood particularly those subjects suffering from hepatic lesion or inflammation [83].

5.7 Antibiotic/antimicrobial

Infectious diseases are becoming the fastest cause of death globally. A number of bacterial etiological agents cause infections, and the use of antibiotics has become a panacea treatment to the ravishing effects of these microbiological agents. However, it is worthy of mention that the use of antibiotics despite their side effects is in recent times becoming ineffective due to the resistance of these microorganisms which is rapidly increasing [20]. In fact, as a result of these unpalatable trend in the antibiotics use, ongoing efforts have embraced the use of MPs in treating infectious ailments, and a number of plants such as ginger are endowed with established antimicrobial effects as reflected in arrays of in vitro, in vivo, and preclinical studies using different solvents of extraction (ethanol, ethyl acetate, hexane) to inhibit microbial growth as presented by Rahmani et al. [20] and Gunathilake and Rupasinghe [34] from many submissions. Ginger derivatives such as 6-dehydrogingerdione, 6-gingerol, 10-gingerol, and 6-shogaol have established antibacterial effects against strains of bacteria and mycobacterial including *Acinetobacter baumannii*, *Helicobacter pylori*, *Mycobacterium avium*, and *M. tuberculosis* [17, 20, 34]. Interestingly, to corroborate the effectiveness of ginger and/or its derivatives, a report of potency surpassing common synthetic antibiotics in the fight against infectious diseases is noted [17, 34, 57].

5.8 Neuroprotective

Neuroprotection refers to the way and manner the central nervous system (CNS) is shielded from neuronal damages resulting from acute and/or chronic neurodegenerative disorders (such as stroke, Alzheimer's, Huntington's, Parkinson's diseases) as a consequence of CNS neurons breakdown and/or worsening of the cognitive or intellectual reasoning of the patients [84]. Intriguingly, the emergence of neurodegenerative diseases (NDD) is age-related, i.e., as individual age, so the possibility of suffering from NDD [85]. Medicinal plants such as ginger have continued to find its place in the management and/or treatment of diseases particularly NDD, and these effects are attributed to its inherent phenolic and flavonoid compounds [17, 20]. A root extract of ginger at 100 mg/kg bw extenuates the effect of monosodium glutamate-induced toxicity in rats (**Table 1**). The emergence or onset of many diseases is triggered by the production of FRs; similarly, since one of the complications of DM is neuropathy, hence, a relationship between FR, NDD, and diabetes is noted. Actually, ginger in separate studies was reported to promote or strengthen the antioxidant defense mechanism of the rat's brain following streptozotocin induction [86–88]. Furthermore, 6-shogaol was studied to inhibit microglia in transient global ischemia [89].

5.9 Antiulcer/gastroprotective

Ulcer (gastric or duodenal) is also a disease affecting majority of the populations of the world for more than ten (10) decades now [90], caused by discrepancies between the protective factors (bicarbonates, prostaglandins, mucin, nitric oxide) and aggressive factors (acid and pepsin) leading to a great deal of mortality and morbidity. Several factors [etiologic (*Helicobacter pylori*) or otherwise, e.g., sedentary lifestyle, diet, drug (NSAIDs), smoking, bacterial infection, free radicals, etc.] influence the emergence and/or progression of ulcer. The treatment involves the use of antimicrobial drugs (metronidazole, tetracycline, amoxicillin, etc.) geared toward eliminating *H. pylori*, antisecretory agents (omeprazole and so on), antagonist of H2 receptors (cimetidine, ranitidine, etc.), and other agents targeting the disruption of the cell wall or membrane of the bacteria (bismuth salt). However, these series of therapies bring about toxicities, thus the clamor for the alternative form of treatment with little or no toxicities, qualities found in medicinal plants such as ginger. The antiulcerative action of ginger is achieved via the elevation of mucin production [20] and enzyme (thromboxane synthetase) inhibition [17]. A number of studies proving the gastroprotective properties of ginger and some of its constituents such as 6-gingerol and 6-shogaol had been established as compiled or presented in the work of Rahmani et al. [20].

5.10 Antiemetic

Ginger in a study using rodents was found to possess anti-serotonin and 5-HT3 receptor antagonism effect in inducing nausea and vomiting during post-surgery [91]. Derivatives of ginger such as gingerol, shogaols, galanolactone, and diterpenoid were also established to reduce nausea and vomiting [92]. Others revealed that the reports of management of nausea and vomiting in cancer patients are also available in the literature [93].

5.11 Hepatoprotective

The liver is the second largest organ (after the skin) in the body where metabolism of drugs or chemical substance occurs. Hence, important attention is required for this organ for good health status and well-being. Liver ailments also constitute a major health problem in the world today caused sometimes by exposure or ingestion of toxic chemicals (carbon tetrachloride, thioacetamide, certain antibiotics, excessive alcohol intake, etc.), and the use of conventional drugs for the treatment of liver diseases is ineffective and comes with side effects. However, solace has been found with MPs such as ginger as alternative means to treating these ailments. Report of relief from liver cirrhosis following carbon tetrachloride-induced liver toxicity in rats as ginger singly or either in combination with curcumin at 100 mg/kg bw ameliorated the liver injury to the animal [94]. Additionally, ginger in another report at 200, 400 mg/kg bw fortified the activity of antioxidants enzymes (superoxide dismutase, catalase, glutathione peroxidase) while lowering the activity of liver function enzymes (alanine transaminase, aspartate aminotransferase) in the acetaminophen-induced hepatic injury [62] as also corroborated by Rahmani et al. [20] in several studies.

5.12 Toxicity profiles

Toxicity may be acute, subacute, chronic, and subchronic [95]. These studies are carried out to provide information about the safety profile of a substance. Medicinal plants are used as a form of therapeutic measure over a long period against numerous diseases. In fact, despite the fact that the active precursors of a number of chemical moieties or drugs are obtained from plant, the acceptance of herbal medicine and/or formulations are exceedingly growing globally. Intriguingly, 80% of the entire global population are using herbal products for the maintenance of their health due to their perceived thought of originating from nature, lesser side effects, efficacy, safety, affordability, etc., although in some quarters, a very few of these medicinal plants have been reported to cause one form of illness (to the liver and kidney). However, report from several studies has not linked ginger in a way to any of these injuries. This fact is corroborated in reports ascertaining the safety of ginger in different concentrations, 0.5–1.0 g, 2.5 g/kg, 100, 333, 500, 1000, 2000 mg/kg bw, in animal studies for different experimental study period ranging from 10 days, 35 days, 3 months to 2 and half years as nontoxic [20] even during pregnancy (rats) and gynecological operation as revealed by a clinical study [34].

5.13 Other pharmacological activities

The effectiveness of ginger against diseases affecting the eye and other ailments such as osteoarthritis, migraine attack, platelet aggregation, gastrointestinal disturbances, nematode invasion, etc. has been established [15, 17, 18, 20, 34].

6. Conclusion

The world is filled with enormous diseases causing major setbacks to the health status of humanity. Unfortunately, the synthetic moieties adopted for therapeutic and preventive measures are not helping (at all) as they are characterized with side effects. Medicinal plants such as ginger are now being embraced as the alternative options for combating various simple or life-threatening ailments. Since various efforts had established the effectiveness of ginger and its corresponding derivatives on a number of ill-health (though lacking clinical reports), there is much hope in the future that ginger might be able to rescue humankind from these evolving derangements causing setbacks to their living and/or survival.

Acknowledgements

The authors acknowledge Directorate Research and Development, University of Free State, South Africa, for the Postdoctoral Research Fellowship granted by Dr. FO Balogun tenable in the research group of Phytomedicine and Phytopharmacology at the Department of Plant Sciences, Faculty of Natural and Agricultural Sciences, University of the Free State, Qwaqwa, Free State.

Author details

Fatai Oladunni Balogun*, Esther Tayo AdeyeOluwa and Anofi Omotayo Tom Ashafa
Phytomedicine and Phytopharmacology Research Group, Faculty of Natural and Agricultural Sciences, Department of Plant Sciences, University of the Free State, Phuthadijthaba, Qwaqwa, Free State, South Africa

*Address all correspondence to: balogunfo@yahoo.co.uk

References

[1] Wakchaure R, Ganguly S. Phytochemistry and pharmacological properties of ginger (*Zingiber officinale*). In: Mahdi AA, Abid M, Khan AA, Ansair MI, Maheshwari RK, editors. Molecular Biology and Pharmacognosy and Beneficial Plants. Delhi: Lenin Media PVT. LTD.; 2018. pp. 97-103

[2] Park EJ, Pizzuto JM. Botanicals in cancer chemoprevention. Cancer and Metastasis Reviews. 2002;**21**:231-255

[3] Rout GR, Das P, Goel S, Raina SN. Determination of genetic stability of micropropagated plants of ginger using random amplified polymorphic DNA (RAPD) markers. Botanical Bulletin- Academia Sinica. 1998;**39**:23-37

[4] Nair KPP. The agronomy and economy of ginger. In: Nair KPP, editor. The Agronomy and Economy of Turmeric and Ginger. Edinburgh, United Kingdom: Elsevier; 2013. pp. 225-292

[5] Ali BH, Blunden G, Tanira MO, Nemmar A. Some phytochemical, pharmacological and toxicological properties of ginger (*Zingiber officinale* Roscoe): A review of recent research. Food and Chemical Toxicology. 2008;**46**:409-420

[6] Sharma PC, Yelne MB, Dennis TJ. Database on Medicinal Plants used in Ayurveda. New Delhi: Central Council for Research in Ayurveda and Siddha, Department of Indian system of medicine. Govt. of India; 2001; 1: 152 2: 177; 4: 90, 213, 404

[7] Morakinyo AO, Oludare GO, Aderinto OT, Tasdup A. Antioxidant and free radical scavenging activities of aqueous and ethanol extracts of *Zingiber officinale*. Biologie et Médecine. 2011;**3**:25-30

[8] Gupta M. Pharmacological properties and traditional therapeutic uses of important Indian spices: A Review. International Journal of Food Properties. 2010;**13**(5):1092-1116. DOI: 10.1080/10942910902963271

[9] Raal A, Volmer D, Sõukand R, Hratkevitš S, Kalle R. Complementary treatment of the common cold and flu with medicinal plants—results from two samples of pharmacy customers in Estonia. PLoS One. 2013;**8**:e58642. DOI: 10.1371/journal.pone.0058642

[10] Khayat S, Kheirkhah M, Behboodi Moghadam Z, Fanaei H, Kasaeian A, Javadimehr M. Effect of treatment with ginger on the severity of premenstrual syndrome symptoms. International Scholarly Research Notices: Obstetrics and Gynecology. 2014;**2014**:79708. DOI: 10.1155/2014/792708

[11] Gull I, Saeed M, Shaukat H, Aslam SM, Samra Z, Athar AM. Inhibitory effect of *Allium sativum* and *Zingiber officinale* extracts on clinically important drug resistant pathogenic bacteria. Annals of Clinical Microbiology and Antimicrobials. 2012;**11**:8. DOI: 10.1186/1476-0711-11-8

[12] Lee SH, Cekanova M, Baek SJ. Multiple mechanisms are involved in 6-gingerol-induced cell growth arrest and apoptosis in human colorectal cancer cells. Molecular Carcinogenesis. 2008;**47**:197-208. DOI: 10.1002/mc.20374

[13] Maxwell I. Let's make ginger beer. Dave's Garden. 2008

[14] Weidner MS, Sigwart K. Investigation of the teratogenic potential of *Zingiber officinale* extract in the rat. Reproductive Toxicology. 2000;**15**:75-80. DOI: 10.1016/S0890-6238(00)00116-7

[15] Rehman R, Akram M, Akhtar N, Jabeen Q, Saeed T, Shah A, et al. *Zingiber officinale* Roscoe (pharmacological activity). Journal of Medicinal Plant Research. 2011;**5**(3):344-348

[16] Pratap SR, Gangadharappa HV, Mruthunjaya K. Ginger: A potential nutraceutical, an updated review. Journal of Pharmacognosy and Phytochemical Research. 2017;**9**(9):1227-1238

[17] Mele MA. Bioactive compounds and biological activity of ginger. Journal of Multidisciplinary Science. 2019;**1**(1):1-7

[18] Gupta SK, Sharma A. Medicinal properties of *Zingiber officinale* Roscoe—A Review. IOSR Journal of Pharmacy and Biological Sciences. 2014;**9**(5):124-129

[19] Sheetal G, Jamuna P. Studies on Indian green leafy vegetables for their antioxidants activity. Plant Foods for Human Nutrition. 2009;**64**:39-45

[20] Rahmani AH, Shabrmi FM, Aly SM. Active ingredients of ginger as potential candidates in the prevention and treatment of diseases via modulation of biological activities. International Journal of Physiology, Pathophysiology and Pharmacology. 2014;**6**:125-136

[21] Prasad S, Tyagi AK. Ginger and its constituents: Role in prevention and treatment of gastrointestinal cancer. Gastroenterology Research and Practice. 2015;**2015**:142979. DOI: 10.1155/2015/142979

[22] Shukla Y, Singh M. Cancer preventive properties of ginger: A brief review. Food and Chemical Toxicology. 2007;**45**(5):683-690

[23] Ugwoke CEC, Nzekwe U. Phytochemistry and proximate composition of ginger (*Zingiber officinale*). Journal of Pharmaceutical and Allied Sciences. 2010;7(5). DOI: 10.4314/jophas.v7i5.63462

[24] Dhanik J, Arya N, Nand V. A review on *Zingiber officinale*. Journal of Pharmacognosy and Phytochemistry. 2017;**6**(3):174-184

[25] Grzanna R, Lindmark L, Frondoza CG. Ginger—an herbal medicinal product with broad anti-inflammatory actions. Journal of Medicinal Food. 2005;**8**(2):125-132

[26] Ashraf K, Sultan S, Shah SAA. Phytochemistry, phytochemical, pharmacological and molecular study of *Zingiber officinale* Roscoe: A review. International Journal of Pharmacy and Pharmaceutical Sciences. 2017;**9**(11):8-16. DOI: 10.22159/ijpps.2017v9i11.19613

[27] Butt MS, Sultan MT. Ginger and its health claims: Molecular aspects. Critical Reviews in Food Science and Nutrition. 2011;**51**(5):383-393. DOI: 10.1080/10408391003624848

[28] Ravindran PN. Ginger (*Zingiber officinale*). In: Jain S, Russel R, editors. Encyclopedia of Herbs and Spices. Vol. 1. Glasgow: Bell and Bain; 2016. pp. 397-409. Available from: https://Iccn.loc.gov/2016029187

[29] Connell DW. The pungent principles of ginger and their importance in certain ginger products. Food Technology. 1969;**21**:570-575

[30] Govindarajan VS. Ginger-chemistry, technology, and quality evaluation. Part 1. Critical Reviews in Food Science and Nutrition. 1982;**17**:1-96

[31] Ahmad B, Rehman MU, Amin I, Arif A, Rasool S, Bhat SA, et al. A review on pharmacological properties of Zingerone (4-(4-Hydroxy-3-methoxyphenyl)-2-butanone). The Scientific World Journal. 2015;**2015**:816364. DOI: 10.1155/2015/816364

[32] Gupta R, Singh PK, Singh R, Singh RL. Pharmacological activities of *Zingiber officinale* (ginger) and its active ingredients: a review. International Journal of Scientific and Innovative Research. 2016;**4**(1):1-18

[33] de Lima RMT, Dos Reis AC, de Menezes APM, Santos JVO, Filho JWGO, Ferreira JRO, et al. Protective and therapeutic potential of ginger (*Zingiber officinale*) extract and [6]-gingerol in cancer: A comprehensive review. Phytotherapy Research. 2018;**32**(10):1885-1907. DOI: 10.1002/ptr.6134

[34] Gunathilake KDPP, Vasantha Rupasinghe HP. Recent perspectives on the medicinal potential of ginger. Botanics: Targets and Therapy. 2015;**5**:55-63

[35] Roufogalis BD. *Zingiber officinale* (Ginger): A future outlook on its potential in prevention and treatment of diabetes and prediabetic states. New Journal of Science. 2014;**2014**:674684. DOI: 10.1155/2014/674684

[36] Chaiyakunapruk N, Kitikannakorn N, Nathisuwan S, Leeprakobboon K, Leelasettagool C. The efficacy of ginger for the prevention of postoperative nausea and vomiting: A meta-analysis. American Journal of Obstetrics and Gynecology. 2006;**194**:95-99

[37] Ali A, Gilani AH. Medicinal value of ginger with focus on its use in nausea and vomiting of pregnancy. International Journal of Food Properties. 2007;**10**(2):269-278. DOI: 10.1080/10942910601045297

[38] Reddy AA, Lokesh BR. Studies on spice principles as antioxidants in the inhibition of lipid peroxidation of rat liver microsomes. Molecular and Cellular Biochemistry. 1992;**111**:117-124

[39] Bardi DA, Halabi MF, Abdullah NA, Rouhollahi E, Hajrezaie M, Abdulla MA. *In vivo* evaluation of ethanolic extract of *Zingiber officinale* rhizomes for its protective effect against liver cirrhosis. BioMed Research International. 2013;**2013**:918460. DOI: 10.1155/2013/918460

[40] Kim JK, Kim Y, Na KM, Surh YJ, Kim TY. [6]-Gingerol prevents UVB-induced ROS production and COX-2 expression *in vitro* and *in vivo*. Free Radical Research. 2007;**41**(5):603-614

[41] Afshari AT, Alireza S, Amirabbas F, Saadatian R, Rasmi Y, Saboory E, et al. The effect of ginger on diabetic nephropathy, plasma antioxidant capacity and lipid peroxidation in rats. Food Chemistry. 2007;**101**(1):148-153

[42] Dugasani S, Pichika MR, Nadarajah VD, Balijepalli MK, Tandra S, Korlakunta JN. Comparative antioxidant and anti-inflammatory effects of [6]-gingerol, [8]-gingerol, [10]-gingerol and [6]-shogaol. Journal of Ethnopharmacology. 2010;**127**:515-520

[43] Li F, Wang Y, Parkin KL, Nitteranon V, Liang J, Yang W, et al. Isolation of quinone reductase (QR) inducing agents from ginger rhizome and their *in vitro* anti-inflammatory activity. Food Research International. 2011;**44**:1597-1603

[44] Bellik Y. Total antioxidant activity and antimicrobial potency of the essential oil and oleoresin of *Zingiber officinale* Roscoe. Asian Pacific Journal of Tropical Disease. 2014;**4**:40-44

[45] Karna P, Chagani S, Gundala SR, Rida PC, Asif G, Sharma V, et al. Benefits of whole ginger extract in prostate cancer. The British Journal of Nutrition. 2012;**107**(4):473-484

[46] Habib SHM, Makpol S, Hamid NAA, Das S, Ngah WZW, Yusof YAM. Ginger extract (*Zingiber officinale*) has anti-cancer and

anti-inflammatory effects on ethionine-induced hepatoma rats. Clinics. 2008;**63**(6):807-813

[47] Akimoto M, Iizuka M, Kanematsu R, Yoshida M, Takenaga K. Anticancer effect of ginger extract against pancreatic cancer cells mainly through reactive oxygen species-mediated autotic cell death. PLoS One. 2015;**10**(5):e0126605

[48] Ojewole JA. Analgesic, anti-inflammatory and hypoglycaemic effects of ethanol extract of *Zingiber officinale* (Roscoe) rhizomes (Zingiberaceae) in mice and rats. Phytotherapy Research. 2006;**20**(9):764-772

[49] Sharma JN, Srivastava KC, Gan EK. Suppressive effects of eugenol and ginger oil on arthritic rats. Pharmacology. 1994;**49**(5):314-318

[50] Abdul Razaq NB, Cho MM, Win NN, Zaman R, Rahman MT. Beneficial effects of ginger (*Zingiber officinale*) on carbohydrate metabolism in streptozotocin-induced diabetic rats. The British Journal of Nutrition. 2012;**108**(7):1194-1201

[51] Sattar NA, Hussain F, Iqbal T, Sheikh MA. Determination of *in vitro* antidiabetic effects of *Zingiber officinale* Roscoe. Brazilian Journal of Pharmaceutical Sciences. 2012;**48**(4):601-607

[52] Sukalingama K, Ganesana K, Ganib SB. Hypoglycemic effect of 6-gingerol, an active principle of ginger in streptozotocin induced diabetic rats. Research and Reviews : Journal of Pharmacology and Toxicological Studies. 2013;**96**:660-666

[53] Ghayur MN, Anwarul HG, Afridi MB, Houghton PJ. Cardiovascular effects of ginger aqueous extract and its phenolic constituents are mediated through multiple pathways. Vascular Pharmacology. 2005;**43**(4):234-241

[54] Thompson M, Al-Qattan KK, Al-Sawan SW, Al-Nageeb MA, Khan I, et al. The use of ginger as a potential anti-inflammatory and antithrombotic agent, Prostagladin, leukotriens and essential fatty acids. Prostaglandins, Leukotrienes & Essential Fatty Acids. 2002;**67**(6):475-478

[55] Malu SP, Obochi GO, Tawo EN, Nyong BE. Antibacterial activity and medicinal properties of ginger (*Zingiber officinale*). Global Journal of Pure and Applied Sciences. 2008;**15**(3):365-368

[56] Ekwenye UN, Elegalam NN. Antibacterial activity of ginger (*Zingiber officinale* Roscoe) and garlic (*Allium sativum* L) extracts on *Escherichia coli* and *Salmonella typhi*. Journal of Molecular Medicine and Advance Sciences. 2005;**1**(4):411-416

[57] Sebiomo A, Awofodu AD, Awosanya AO, Awotona FE, Ajayi AJ. Comparative studies of antibacterial effect of some antibiotics and ginger (*Zingiber officinale*) on two pathogenic bacteria. Journal of Microbiology and Antimicrobials. 2011;**3**(1):18-22

[58] Waggas AM. Neuroprotective evaluation of extract of ginger (*Zingiber officinale*) root in monosodium glutamate induced toxicity in different brain areas male albino rats. Pakistan Journal of Biological Sciences. 2009;**12**(3):201-212

[59] Johji Y, Michihiko M, Rong HQ, Hisashi M, Hajime F. The anti-ulcer effect in rats of ginger constituents. Journal of Ethnopharmacology. 1988;**23**(2-3):299-304

[60] Suekawa M, Ishige A, Yuasa K, Sudo K, Aburada M, Hosoya E. Pharmacological studies on ginger. I. Pharmacological actions of pungent constituents, (6)-gingerol and (6)-shogaol. Journal of Pharmacobio-Dynamics. 1984;7:13-18

[61] Suekawa M, Ishige A, Yuasa K, Sudo K, Aburada M, Hosoya E. Pharmacological

studies on ginger. I. Pharmacological actions of pungent constituents, (6)-gingerol and (6)-shogaol. Journal of Pharmacobio-Dynamics. 1984;7(11):836-848

[62] Ajith TA, Hema U, Aswathy MS. *Zingiber officinale* Roscoe prevents acetaminophen-induced acute hepatotoxicity by enhancing hepatic antioxidant status. Food and Chemical Toxicology. 2007;**45**:2267-2272

[63] Krishnakantha TP, Lokesh BR. Scavenging of superoxide anions by spice principles. Indian Journal of Biochemistry & Biophysics. 1993;**30**:133-134

[64] Cao ZF, Chen ZG, Guo P, Zhang SM, Lian LX, Luo L, et al. Scavenging effects of ginger on superoxide anion and hydroxyl radical. Chung Kuo Chung Yao Tsa Chih. 1993;**8**:750-764

[65] Stoilova I, Krastanov A, Stoyanova A, Denev P, Gargova S. Antioxidant activity of a ginger extract (*Zingiber officinale*). Food Chemistry. 2007;**102**(3):764-770

[66] Sueishi Y, Masamoto H, Kotake Y. Heat treatments of ginger root modify but not diminish its antioxidant activity as measured with multiple free radical scavenging (MULTIS) method. Journal of Clinical Biochemistry and Nutrition. 2019;**64**(2):143-147

[67] Greenwell M, Rahman PKSM. Medicinal plants: Their use in anticancer treatment. International Journal of Pharmaceutical Sciences and Research. 2015;**6**(10):4103-4112. DOI: 10.13040/IJPSR.0975-8232.6(10).4103-12

[68] Vemuri SK, Banala RR, Subbaiah GPV, Srivastava SK, Reddy AG, Malarvili T. Anti-cancer potential of a mix of natural extracts of turmeric, ginger and garlic: A cell-based study. Egyptian Journal of Basic and Applied Sciences. 2017;**4**(4):332-344

[69] Ghasemian M, Owlia S, Owlia MB. Review of anti-inflammatory herbal medicines. Advances in Pharmacological Sciences. 2016;**2016**:9130979. DOI: 10.1155/2016/9130979

[70] Rani PM, Padmakumari KP, Sankarikutty B, Lijo Cherian O, Nisha VM, Raghu KG. Inhibitory potential of ginger extracts against enzymes linked to type 2 diabetes, inflammation and induced oxidative stress. International Journal of Food Sciences and Nutrition. 2011;**62**(2):106-110

[71] Parmar I, Rupasinghe HPV. Antioxidant capacity and anti-diabetic activity of wild berry stem infusions. European Journal of Medicinal Plants. 2015;**8**:11-28

[72] Balogun FO, Ashafa AOT. A review of plants in South African traditional medicine used in the prevention and management of hypertension. Planta Medica. 2019;**85**:312-334

[73] Aming SN. The effect of twice a day intake of ginger tea on the blood pressure of hypertensive individuals in Barangay La Victoria, Aurora, Zamboanga Del Sur. Herdin Record #: R09-ZCHRD-12043023205433. 2006 (abstract)

[74] Kuppusamy P, David RS, Raj P, Ilavenil S, Kaleeswaran B, Govindan N, et al. Evaluation of antihypercholesterolemic effect using *Memecylon edule* Roxb. ethanolic extract in cholesterol-induced Swiss albino mice. Journal of Acute Medicine. 2015;**5**:85-e91

[75] Iffiu-Soltesz Z, Wanecq E, Lomba A, Portilio MP, Pellati F, Szoto E. Chronic benzylamine administration in the drinking water improves glucose tolerance, reduces body weight gain and circulating cholesterol in high-fat diet-fed mice. Pharmacological Research. 2010;**61**:355-363

[76] Huseini HF, Kianbakht S, Hajiaghaee R, Dabaghian FH. Antihyperglycemic and anti-hyper-cholesterolemic effects of *Aloe vera* leaf gel in hyperlipidemic type 2 diabetic patients: a randomized double-blind placebo-controlled clinical trial. Planta Medica. 2012;**78**:311-316

[77] Balogun FO, Ashafa AOT. Aqueous roots extract of *Dicoma anomala* (Sond.) ameliorates isoproterenol – induced myocardial infarction in Wistar rats. Tropical Journal of Pharmaceutical Research. 2016a;**15**(8):1651-1657

[78] Balogun FO, Ashafa AOT. Protective action of aqueous leaf extract of *Gazania krebsiana* (Less.) 'Asteraceae' antagonizes isoproterenol-triggered myocardial infarction in *Rattus norvegicus*. Comparative Clinical Pathology. 2018;**27**:461-470

[79] Duke JA. Handbook of Medicinal Herbs. Maryland, USA: CRC Press; 2002

[80] Thomson M, Al-Qattan KK, Al-Sawan SM, Alnaqeeb MA, Khan I, Ali M. The use of ginger (*Zingiber officinale* Rosc) as a potential antiinflammatory and antithrombotic agent. Prostaglandins, Leukotrienes, and Essential Fatty Acids. 2002;**67**(6):475-478

[81] Bhandari U, Kanojia R, Pillai KK. Effect of ethanolic extract of *Zingiber officinale* on dyslipidaemia in diabetic rats. Journal of Ethnopharmacology. 2005;**97**(2):227-230

[82] Nammi S, Sreemantula S, Roufogalis BD. Protective effects of ethanolic extract of *Zingiber officinale* rhizome on the development of metabolic syndrome in high-fat diet-fed rats. Basic & Clinical Pharmacology & Toxicology. 2009;**104**(5):366-373

[83] Heeba GH, Abd-Elghany MI. Effect of combined administration of ginger (*Zingiber officinale*) and atorvastatin on the liver of rats. Phytomedicine. 2010;**17**(14):1076-1081

[84] Elufioye TO, Berida TI, Habtemariam S. Plants-derived neuroprotective agents: Cutting the cycle of cell death through multiple mechanisms. Evidence-based Complementary and Alternative Medicine. 2017;**2017**:3574012. DOI: 10.1155/2017/3574012

[85] Uddin R, Kim HH, Lee J, Park SU. Neuroprotective effects of medicinal plants. EXCLI Journal. 2013;**12**:541-545

[86] Shanmugam KR, Mallikarjuna K, Kesireddy N, Sathyavelu RK. Neuroprotective effect of ginger on anti-oxidant enzymes in streptozotocin-induced diabetic rats. Food and Chemical Toxicology. 2011;**49**:893-897

[87] Sharma P, Singh R. Neuroprotective effect of ginger juice against dichlorvos and lindane induced toxicity in wistar rats. Planta Medica. 2011;**77**:122

[88] El-Akabawy G, El-Kholy W. Neuroprotective effect of ginger in the brain of streptozotocin-induced diabetic rats. Annals of Anatomy. 2014;**196**(2-3):119-128

[89] Ha SK, Moon E, Ju MS, Kim DH, Ryu JH, Oh MS, et al. 6-Shogaol, a ginger product, modulates neuroinflammation: A new approach to neuroprotection. Neuropharmacology. 2012;**63**(2):211-223

[90] Asnaashari S, Dastmalchi S, Javadzadeh Y. Gastroprotective effects of herbal medicines (roots). International Journal of Food Properties. 2018;**21**(1):902-920

[91] Vutyavanich T, Kraisarin T, Ruangsri RA. Ginger for nausea and vomiting in pregnancy: Randomized, double-masked, placebo-controlled

trial. Obstetrics and Gynecology. 2001;**97**(4):577-582

[92] Bhattarai S, Tran VH, Duke CC. The stability of gingerol and shogaol in aqueous solutions. Journal of Pharmaceutical Sciences. 2001;**90**(10):1658-1664

[93] Revol B, Gautier-Veyret E, Arrivé C, Fouilhé Sam-Laï N, McLeer-Florin A, Pluchart H, et al. Pharmacokinetic herb-drug interaction between ginger and crizotinib. British Journal of Clinical Pharmacology. 2019:1-2

[94] Abd-Allah GA, El-Bakry KA, Bahnasawy MH, El-Khodary ER. Protective effects of curcumin and ginger on liver cirrhosis induced by carbon tetrachloride in rats. International Journal of Pharmacology. 2016;**12**:361-369

[95] Balogun FO, Ashafa AOT. Acute and sub-chronic oral toxicity evaluation of aqueous roots extract of *Dicoma anomala* (Sond.) in Wistar rats. Evidence-based Complementary and Alternative Medicine. 2016b;**2016**:3509323. DOI: 10.1155/2016/3509323